BIOMEDICINE

Edited by **Chao Lin**

Biomedicine
Edited by Chao Lin

Published by InTech
Janeza Trdine 9, 51000 Rijeka, Croatia

Publishing Process Manager Anja Filipovic

Technical Editor Teodora Smiljanic

Cover Designer InTech Design Team

Additional hard copies can be obtained from orders@intechopen.com

Biomedicine, Edited by Chao Lin
p. cm.
ISBN 978-953-51-0352-3

For EU product safety concerns:
IN TECH d.o.o., Prolaz Marije Krucifikse Kozulić 3, 51000 Rijeka, Croatia,
info@intechopen.com or visit our website at intechopen.com.

Meet the editor

Dr Chao Lin is an associate professor and research director in nanomedicine at the Institute for Advanced Materials and Nanobiomedicine, affiliated to the School of Medicine, Tongji University (Shanghai, PR China). Before joining the institution, he was a postdoctoral fellow at MIRA-Institute for Biomedical Technology and Technical Medicine, University of Twente (Enschede, The Netherland). He received his PhD from the University of Twente in 2008 under the supervision of Professor Jan Feijen and Professor Johan F.J. Engbersen. He has authored 30 research articles in peer-reviewed journals and 2 book chapters. He serves as an editorial board member of the Journal of Biomaterials and Tissue Engineering. In 2007 he received the Chinese Government Award for Outstanding Self-financed Students Abroad.

Contents

Preface

How to successfully circumvent human disease? How to efficaciously treat human disease? How to healthily live to a decent age? These problems have baffled our ancestors for a few centuries. People have attempted to address these puzzled problems since recorded time, but the answers are often disruptive and even unreliable in part due to the modification or alteration of medical principles in every history era. Over the past decades, comprehensive and systematic scientific researches in cell biology and molecular biology have accelerated the evolution of traditional medicine. Scientists are now able to investigate these problems on the basis of molecular levels. Thus, the purpose of this book is to interpret clinical and fundamental medical findings through the lens of 21st-century biomedicine. The contents of this book provide readers with a broad spectrum of basic biomedical understanding dealing with regenerative medicine, gene medicine, and medical devices.

Biomedicine can be divided into four sections: regenerative medicine in Chapter 1 and 2; gene medicine and nanobiomedicine in Chapter 3 to 5; biomedical device performance in Chapter 6 and 7; and public perception of medicine in Chapter 8. Each chapter is intently chosen and written by invited experts or physicians from professional biomedical field, to address hot issues related with contemporary biomedical science. Additionally, the core concepts in each chapter are highlighted and the authors wish readers to grasp them smoothly through readily readable schemes and colourful illustrations. The reference lists also comprise state-of-the-art reviews, relevant research articles, proceedings in frontier conferences, which are certainly helpful to beginners for further learning.

Chapter 1 provides a complete overview on current methods for the protection of isolated pancreatic islets from host immune response. The authors emphasize an emerging lay-by-lay technique as an innovative route to maintain viability and functions of islet, implying great promise for the treat of diabetes. Next, in-situ forming hydrogels as scaffolds for tissue regeneration are summarized in Chapter 2. Several types of hydrogels that are bioactive and bio-response are outlined in detail and indicated to be highly desirable for clinical practices. Chapter 3 gives an introduction on recent progress in RNA interference (RNAi) for cancer therapy, including mechanisms of gene silencing, cancer-related gene targets for RNAi and

RNAi-based clinical trails, although there remain big challenges for clinical RNAi. A latest research arena in the improvement of clinical drug availability is the "nanobiomedicine", since nanoscale carriers loaded with drugs or genes reveal high possibility for efficient and targeted delivery. Chapter 4 presents the advancement in disulfide-based cationic polymers, in response to intracellular reducing milieu, as nano-carriers for gene transfection. Chapter 5 brings a new outlook on stable magnetic isotopes, which might offer the generation of new pharmaceuticals for biomedical applications. Medical devices are very crucial for diagnosis and therapy of human diseases. Chapter 6 provides recent research progress of fiber-gating sensors for medical diagnosis to facilitate surgery and clinical therapy. The history, working principle and fabrications of optical fiber gratings are introduced thoroughly. In Chapter 7, electron beam melting for processing ferrous metallic powders to fully-dense materials is discussed and this technique is valuable in the field of medical implants. At last, a review in Chapter 8 points out the evolution process of public perception on biomedicine in the last decades through survey and statistical data in Spain and other European countries. This chapter is constructive to alarm scientists to take into account public awareness on medicine rather than scientific activity alone.

Thus, I truly believe that *Biomedicine* is a valuable book that may assist readers to be aware of current advancement in rapidly moving field of biomedicine.

Chao Lin
Tongji University School of Medicine
Tongji University
Shanghai,
PR China

Part 1

Regenerative Medicine

1

Encapsulation and Surface Engineering of Pancreatic Islets: Advances and Challenges

Veronika Kozlovskaya, Oleksandra Zavgorodnya and Eugenia Kharlampieva
Department of Chemistry, University of Alabama at Birmingham USA

1. Introduction

Type 1 diabetes (T1D) is a chronic autoimmune disease representing a major health care problem worldwide (Tierney et al., 2002). T1D is caused by islet-reactive immune T cells that destroy insulin-producing pancreatic β-cells. Transplantation of insulin-producing pancreatic islets by their injection in vascularized organs has been recently recognized as a promising path to curing diabetes (Meloche, 2007; Robertson, 2000). However, despite the significant promise, the clinical application of the procedure remains limited due to (a) limited supply of islets suitable for transplantation, (b) a hypoxia because of a low tension of oxygen at the implantation sites and (c) an acute rejection during transplantation. One of the challenges is associated with isolation and culturing islets *in vitro* before injection. In the pancreas, endocrine cells of the islet clusters are separated from exocrine cells by a discontinuous mantle of collagen fibers defining their respective basement membrane. During collagenase isolation of islets from the pancreas, further disruption of the islet mantle results in preparations exhibiting various morphological changes (islet fragmentation, fusion) under routine tissue culture conditions, particularly in human islets (Lacy & Kostianovsky, 1967). Attenuation of islet viability and functionality accompanies these morphological changes. The second issue is associated with islet transplantation which requires immunosuppression to protect the donor islets from the host immune response and prevent implant rejection and post-surgery inflammations (Ricordi & Strom, 2004). Despite the fact that a range of immunosuppressive drugs have demonstrated pharmacologically inhibitory effects on pro-inflammatory cytokines (Riachy et al., 2002; Contreras et al., 2002; Lv et al., 2008; Stosic-Grujicic et al., 2001), the use of immunosuppressive molecules is very specific since they can induce non-specific suppression of the immune system resulting in serious side effects and increased risk of infection which can work against the benefits of a transplant (Narang & Mahato, 2006). These issues have inspired the development of a number of strategies to prevent immunogenic reactions and stabilize islet morphology and functionality, both *in vitro* and following transplantation *in vivo* (Chandy et al., 1999; Abalovich et al., 2001). Two major approaches have been introduced to prevent immunogenic reactions on the islet surfaces: macro and microencapsulation of the islet cells and islet cell surface modification (Fig. 1) (De Vos et al., 2003; Panza et al., 2000; Scott & Murad, 1998; Opara et al., 2010).

Islet macro/microencapsulation strategy is based on embedding islets in solid matrices, allowing for the creation of a semi-permeable environment around islets capable of immune-protection and for mass and oxygen transfer (Beck et al., 2007; Weber et al., 2007). For that, the isolated islets are usually entrapped individually or as islet clusters in thick gels, for example, high-viscous alginate droplets stabilized with divalent ions of barium or calcium (Zimmermann et al., 2001). Islet surface modification strategy involves covalent conjugation of molecules to islet cell surfaces. However, this technology is limited to the introduction of specified functional small molecules to cells and might interfere with cell physiology (Rabuka et al., 2008; Paulick et al., 2007). Layer-by-layer (LbL) technique has been recently applied as a new approach to modify islet surfaces (Krol et al., 2006; Wilson et al., 2008). The technique is based on alternating LbL deposition of water soluble polymers on surfaces from aqueous solutions which results in nano-thin coatings of controllable thickness and composition (Decher & Schlenoff, 2002; Kharlampieva & Sukhishvili, 2006; Tang et al., 2006).

Unlike bulk encapsulating materials, the ultrathin conformal coating affords a faster response to stimulation and the possibility to bind factors or protective molecules to the protective ultrathin shell with the later slow triggered release of these molecules (Chluba et al., 2001). By selecting specific pairs of polyelectrolytes, a defined cutoff of the coating (Kozlovskaya & Sukhishvili, 2006) is possible, as is inhibitor binding to prevent graft rejection, microphage attacks, or antibody recognition (Kim & Park, 2006). Here, we review methods and devices designed for protecting isolated islets from host immune responses while allowing transport of essential nutrients. We also discuss challenges of various approaches developed for encapsulation of individual islets in thin coatings that conform to the islet surfaces, fabricated using a number of physical and chemical processes.

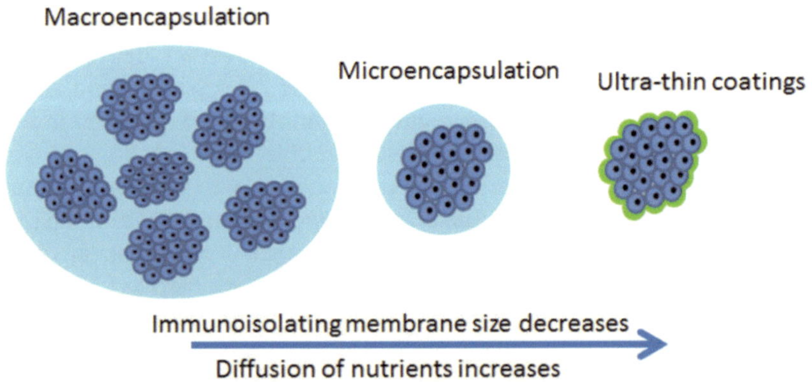

Fig. 1. Strategies for encapsulation and surface engineering of pancreatic islets.

2. Preservation of islets *in vitro*

Recently, the advantages of cultured islets before transplantation have been demonstrated over freshly isolated islets (Herring et al., 2004; Ichii et al., 2007). It is known that cellular stress due to pancreas preservation and islet isolation process leads to a loss of islets during the first 24 h after isolation. Islet culture after isolation can prevent islet cells from toxic factors generated by cells damaged during these processes, providing sufficient oxygen and nutrients to allow islet cells to recover. After isolation of islets from donors, it is crucial to maintain islet viability and functionality until transplantation to give sufficient time to perform microbiological tests as well as donor matching and recipient pre-conditioning.

Modifying the islet preparations for reducing immunogenicity by altering temperature (Kim et al., 2005; Stein et al., 1994), or media composition is one of the advantages for islet pre-culture (Ricordi et al., 1987; Murdoch et al., 2004). For example, supplementation of culture media with lactogen hormones has been shown to minimize β-cell loss during pre-transplant culture leading to a higher β-cell survival rather than proliferation (Yamamoto et al., 2010; Nielsen, 1982). When islets were cultured in media supplied with recombinant human prolactin (rhPRL) for 48 h, production of interferon-gamma (IFN-γ), tissue necrosis factor-alpha (TNF-α), interleukins cytokines, IL-6, IL-8 and microphage inflammatory protein-1-β was comparable with the control group of islets with no increase in pro-inflammatory mediators in the presence of rhPRL suggesting no elevated immunogenicity. Furthermore, the PRL treatment of islet preparations resulted in decreased apoptosis in β-cell subsets, suggesting β-cell specific anti-apoptotic effects of rhPRL (Yamamoto et al., 2010). Another possible issue with the pre-cultured islets is the possibility of islet fusion during incubation, which may lead to hypoxia and starving of the cells. Those result in central necrosis of fused islet aggregates causing a significant loss of islet potency and viability (Ichii et al., 2007).

Apoptosis of human islets after isolation from supporting extracellular matrix is a very common cell pathway *in vitro*. During the first steps towards apoptosis integrin expression is diminished and, consequently, phenotype characteristics are lost and islets stop secreting insulin (Ris et al., 2002). Exploring the parameters important for preventing pre-apoptotic events should help in preserving islet viability and function for long periods of time. The effects of two types of collagen, type I and type IV, and fibronectin, proteins that are generally present in the cell-supporting matrix have been explored (Daoud et al., 2010). Islets have a tendency to spread and form a monolayer on surfaces *in vitro*. The islet monolayer can still be viable without preserving the phenotype characteristics, however, the normal insulin secretion of islets will be lost. Daoud et al. showed that integrity and insulin production of islets can be preserved by presence of fibronectin in the medium (Daoud et al., 2010). Both types of collagen increased the viability of islets from 24 to 48 hours *in vitro*. Several studies revealed an increasing survival of islets *in vitro* when embedded in a solid matrix. Culture in collagen I gels obtained from rat tail and fibrin gels have shown promising for prolonging islet survival (Wang & Rosenberg, 1999; Beattie et al., 2002).

3. Approaches to prevent immunogenic reactions on the islet surfaces

The immune reactions against encapsulated islets can be divided into non-specific immune activation initiated by surgery; a host response against the encapsulating materials and implanted islets provokes the immune response by releasing the bioactive molecules.

The instant blood-mediated inflammatory reaction (IBMIR) is an inflammatory reaction that occurs when isolated islets come in contact with human blood. This process is responsible for islets destruction together with overall failing of transplantation. One of the major triggers for the IBMIR reaction is a secretion of tissue factors by islets surfaces. Disruption of host tissue by surgery leads to release of bioactive molecules such as fibrinogen, histamine and fibronectin. Release of serum components and presence of extracellular matrix and cell debris attract tissue macrophages to the surgery site to clean up and start the process of wound healing. The immune cells have the ability to produce various small bioactive compounds such as interleukins, tissue necrosis factors (TNF) and histamine. There are several extravascular approaches to prevent IBMIR reaction against pancreatic islets (Nafea et al., 2011).

3.1 Macro and microencapsulation

'Macroencapsulation' can be defined as encapsulation of large numbers of islets together in one device. The shape of the devices can be a hollow fiber, planar membranes or macrocapsules. Macroencapsulation provides immune-isolation of islets within semi-permeable membranes. The major advantage of macrocapsules is the possibility to easily retrieve the islets from an implantation site in case of surgery complications. Capsules reduce the risk of the IBMIR reaction from occuring, therefore, allowing the use of not only allogenic but also xenogenic materials and avoiding the use of immunosuppressive medications. However, macroencapsulation has not found a broad use in islet encapsulation and transplantation due to a large volume of protective devices. Relatively large sizes will cause limited oxygen and nutrition access and, as a result, cell necrosis (De Vos et al., 1999). Such macrocapsules do not allow for tuning the molecular weight cutoff (or semi-permeable properties) to prevent recognition by antibodies, and cytokines cannot be sufficiently excluded either (Cui et al., 2004). Similarly, recently developed poly(ethylene glycol) (PEG) hydrogels although demonstrated facile control over porosity but formed microbeads are large and present a barrier for rapid molecular transport (Weber et al., 2007). The capsules with larger diameters than an islet itself are also expected to plug blood vessels. This can exert harmful effects on the patient's liver. The diameter of encapsulated islets must be much smaller than that currently attained to allow transplantation of the islets into portal veins. Thus, new methods for the microencapsulation of islets without increasing the diameter of the implant are required. In this respect, modification of islet surfaces would be a powerful tool that can provide an artificial nurturing environment and preserve islet viability and function (Raymond et al., 2004; Wilson & Chaikof, 2008; Ricordi & Strom, 2004; Lim et al., 2011).

In contrast to macroencapsulation, 'microencapsulation' can be defined as encapsulation of single islets or a small group of islets inside the polymer gel coating. Microencapsulation can be achieved via the formation of a gel shell around the islets by polymerization of a precursor solution around islet surfaces. Emulsification is one of the microbead formation methods (Iwata et al., 1992; Yang et al., 1994). PEG/Alginate was used for islets microencapsulation through their mixing with two-phase aqueous emulsion. The islets contained in emulsion microdroplets underwent cross-linking with calcium ions (Calafiore et al., 2006).

During the microencapsulation process, a bio-inert coating with minimal host response and cell toxicity should be created. Individual islet coating offers a number of advantages over the macroencapsulation. An individual coating provides a better surface:volume ratio that allows faster diffusion of oxygen and nutrients which supports the viability of encapsulated islets. Another advantage is the possibility to employ different coating techniques. The major requirement for the materials used for islets encapsulation is biocompatibility which can be evaluated by a degree of fibrotic overgrowth (Liu et al., 2010). Overgrowth of fibrous tissue upon microcapsule surface would affect the oxygen and nutrition diffusion by clogging microcapsule pores (Nafea et al., 2011). Indeed, insufficient supply of oxygen is the major reason for necrosis of microencapsulated islets. The biocompatibility of the materials used for microcapsules formation strongly depends on chemical composition of materials and applied purification techniques.

Also, biocompatible materials must have selective permeability to promote the survival of encapsulated islets. Selective permeability should allow fluxes of oxygen, nutrients and metabolism products freely in and out of the protective membrane. At the same time the microcoating should prevent the immune system compounds such as antibodies and cytokines to reach the encapsulated cells. The viability and functioning of encapsulated islets is in majority determined by the molecular weight cutoff of the microcapsules. The cutoff determines the upper size limit of molecular weight that is allowed to go through coating. However, diffusion of the molecules very often depends on size, shape and charge of the molecules. Diffusion coefficient and permeability can be considered the more useful and informative characteristics of the cutoff. Control over the permeability can be achieved by varying polymer molecular weight, increasing or decreasing functional group and cross-linking densities and polymerization conditions (Dembczynski & Jankowski, 2001). Molecular weight and concentration of polymer can drastically change the permeability of the hydrogel. It has been shown that poly(vinyl alcohol) (PVA) hydrogel with a high molecular weight of PVA backbone had a higher swelling ratio due to the decrease in cross-link density (Martens et al., 2007). In radical polymerization, an increase in polymer concentration leads to the increase in cross-link density and decrease in the swelling ability. The chemical structure of hydrogel has a significant effect on hydrogel permeability. Presence of electrostatic interactions between charged hydrogel groups and small molecules will slow down the diffusion rates over time. Moreover, during time, changes in the diffusion coefficient occur because of changes in the hydrogel network due to physical or chemical absorption of proteins.

The microcapsule shape (Sakai et al., 2006) morphology such as roughness (Bunger et al., 2003), mechanical properties and especially stiffness of the hydrogel (Berg et al., 2004), play an important role in provoking immune response or fibrotic overgrowths. The importance of hydrogel roughness was demonstrated for implanted rough poly(L-lysine) (PLL)/Alginate microcapsules and only smoothing the surface by adding poly(acrylic acid) completely abolished the tissue response (Bünger et al., 2003). In most cases, smooth round surfaces had the lowest fibrosis promoting effect (Zhang et al., 2008). The most commonly applied materials for microencapsulation are alginate (Lim et al., 1980), agarose (Iwata et al., 1992), PEG (Weber & Anseth, 2008) and poly(hydroxyethylmetacrylate-methyl methacrylate) (Dawson et al., 1987).

3.1.1 Alginate microgels

Alginate microcapsule production can be made under physiological conditions and provide an environment that allows maintaining the islets functionality and viability.

The microcapsule fabrication technique is based on the entrapment of islets within the spherical droplets that are produced by extrusion of solution containing polymer and islets through a needle. Two forces are usually used to control the size of droplets: an air flow that builds around the tip of the needle (Wolters et al., 1992), and a high voltage pulse around the needle tip (Halle et al., 1994; Hsu et al., 1994), the formed droplets very often require an additional stabilization via gelation and beads formation (Stabler et al., 2001). Cross-linking with metal ions, calcium and barium, and chemical or covalent cross-linking are the two methods generally used in microbead or microcapsule formation. Calcium cross-linked alginate microcapsules very often require an additional stabilization with PLL.

Alginate is a linear polysaccharide extracted from algae with a chemical structure combining different ratios of α-L-guluronic (G) and β-D-mannuronic acids (M) and saccharides (Haug et al., 1974). Chemical composition of alginate may affect the biocompatibility (De Vos et al., 1997) or/and function, and the activity of incorporated cells (Stabler et al., 2001). Guluronic acids and alternating mannuronic-guluronic acids have a high ability to cooperatively bind with divalent metal ions and form a cross-linked gel. Barium ion cross-linked gels have a higher stability compared to calcium ion cross-linked gels, though barium ions produce a more stable cross-linking only for alginate with a high ratio of guluronic acid (> 60%). However, ionically cross-linked alginate hydrogels undergo slow degradation under physiological conditions. During this process, alginate microcapsules with metal ion cross-linking undergo slow exchange of divalent cations with sodium ions, which leads to the microgel degradation. Leakage of barium ions is not desirable due to their toxicity which is based on their ability to inhibit K^+ channels at concentration greater than 5-10 mM (Zimmermann et al., 2000). Variations in alginate structures obtained from different alginate sources introduce an additional limitation for the use of the barium cations for alginate cross-linking. Permeability and swelling of alginate-based capsules also strongly depend on the ratio of the acids in alginate. Barium cross-linked G-rich alginate beads had lower permeability to IgG than Ca^{2+} cross-linked microgel. However, cross-linking M-rich alginate beads with Ba^{2+} ions leads to the permeability increase and an overall higher gel swelling (Mørch et al., 2006). Applying an outer coating of polycations such as PLL, or poly-L-ornithine can readily stabilize the alginate capsules with low G-ratio (Thu et al., 1996). Thus, the alginate must be chosen according to a specific application. Moreover, polycations, such as PLL, are proinflammatory molecules responsible for the fibrotic overgrowth (King et al., 2001) and the soluble PLL induces the cytokine production in monocytes and can be a reason for cellular necrosis (Strand et al., 2001).

3.1.2 Poly(ethylene glycol)-based gels

PEG is a less immunogenic material that is generally studied for use in islet encapsulation. PEG is a non-ionic hydrophilic polymer stable at physiological conditions with the highly hydrated polymer coils. The variation in its molecular weight can be used to control protein adsorption and the permeability of PEG gels by changing their porosity unlike alginate (Chen et al., 1998). However, the formed microbeads are larger and diffusion of molecules is slower (Weber et al., 2007a).

Selective-withdrawal coating technique was used to encapsulate rat pancreatic islets within PEG thin coating (Wyman et al., 2007). Geometry of the capsules is determined by selective withdrawal throughout the bulk solution and allows the cross-linking throughout the volume of the formed capsules. Mixture, containing PEG diacrylate, photoinitator accelerator and islets were mixed in a chamber. Once the islets were trapped into droplets, they were drawn into the spout. PEG polymerization was done by the exposure of the droplets to a green light. The free-radical polymerization of PEG diacrylate produce branched polymer chains. The encapsulated islets passed through the withdrawing tube with the same inner diameter that allowed for the same size capsules. Islets encapsulation required two encapsulation rounds. Despite this double coating produced capsules with mean thickness of 20.5 µm, islets were capable to respond to high glucose stimulation without a delay.

Hydrogel network diffusion properties can be adjusted by altering the cross-link density or polymer molecular weight. The PEG microgel cross-linked by photoinitiated polymerization is a well studied system for islet encapsulation. An average pore size of the PEG gel formed by chain polymerization is dependent on the length of the polymer chains between cross-links and can be used to predict and calculate the diffusion coefficient within the gel. Studies revealed that the molecular weight of PEG can not only affect the cross-link density, but also can change the concentration profile of radicals in time. Changes in the concentration of free radicals affect both a number of cross-linkable double bonds and diffusion limitation introduced during gel formation (Weber et al., 2008). More importantly, microcapsules produced from PEG with M_w 2000–10,000 g/mol via photopolymerization did not affect the viability of encapsulated islets or alter their insulin secretion. The level of glucose secretion observed during 1 hour of glucose stimulation was at the same level as for unmodified islets. Insulin secretion from islets encapsulated into 1 mm thick PEG gel was delayed at specific time point within stimulation period. The diffusion-related delay can be minimized through reducing the distance between encapsulated islets and surrounding environment by reducing the thickness of the gel (Weber et al., 2008).

The created hydrogel environment is extremely different from the natural surroundings of islets. Indeed, cell viability is strongly dependent on cell-cell or cell-matrix interactions. For example, the viability of MIN6 pancreatic β-cells encapsulated in just PEG-based matrix reduced to 17% within 10 days (Weber et al., 2007b). Mimicking the extracellular environment in an artificial matrix can provide additional enhancement of islets viability and functioning. For that, adhesive peptides or various biological components for promoting cell-matrix interactions can be co-polymerized within hydrogels. Thus, physical incorporation of laminin and collagen type IV, the most abundant proteins present in the extracellular matrix, was performed during formation of hydrogel from PEG dimethacrylate block co-polymers via photoinitiated polymerization (Weber and Anseth, 2008; Weber et al., 2006). Such introduction of the proteins to the PEG matrix preserved the viability and functioning of the microencapsulated islets up to 28 days in culture. Interestingly, the ratio of the proteins had an effect on the level of insulin secretion from the encapsulated islets. For example, equal amounts of laminin and collagen IV in the hydrogel did not have any synergetic effect on the insulin production, while enhanced insulin secretion was observed for the increasing amount of laminin.

Isolation and purification of matrix proteins is, however, complex and cost inefficient way to recreate extracellular matrix environment. The substitution of natural proteins by artificial sequences is a promising approach to resolve this issue. Recently, several therapeutic agents (Drucker, 2001; Drucker, 2002; Holz & Chepurny, 2003) have been proposed to increase insulinotropic effect of pancreatic cells and protect them from apoptosis. The most promising agent is glucagon-like peptide 1 (GLP-1) that has an extremely short half-life time (less than 2 min) *in vivo* (Lee et al., 2005). Synthesis of artificial peptides with the same biological activities as a substitution for the natural peptide can be one of the solutions for this problem. Synthesis and evaluation of the synthetic glucagon like peptide functionality had been made by Lin et al (Lin & Anseth, 2009). The biological activity of the synthetic peptide physically entrapped in photopolymerized PEG microgel was not affected by photopolymerization. Presence of the synthetic substitute of GLP-1 in the hydrogel had a positive effect on insulin secretion and viability of rat islets encapsulated into the PEG hydrogel.

Many cytotoxic molecules of a low molecular weight are able to easily diffuse inside the microgel capsules *in vivo*. Early islet graft failing and islet damaging occurs due to infiltration of pro-inflammatory cytokines, such as interleukin-1β (IL-1β), tissue necrosis factor-alfa (TNF-α), and interferon-gamma (IFN-γ). The size of these molecules allows them to freely diffuse within the matrixes commonly used in islets encapsulation. A reasonable approach to resolve this issue would be a further reduction of microcapsule permeability. However, the microcapsule pore size cannot be reduced much and still has to allow the permeability of insulin. The decrease in capsule size generally leads to a higher permeability towards cytotoxic molecules. For example, the islets encapsulated into 400-500 µm capsules had minor damages from cytokines, while the islets encapsulated into capsules with smaller dimensions become more susceptible to damage by cytokines (Basta et al., 2004). Moreover, a combination of cytokines and their concentration are also important factors (King et al., 2000; De Vos et al., 2003a).

Incorporation of molecules that can locally inhibit interactions between the small cytotoxic molecules and islets can be used to increase the life-time of encapsulated islets. Indeed, when anti-inflammatory peptide, the cytokine-inhibitory peptide IL-1RIP, had been introduced into PEG hydrogel even in low molar ratio, the survival of islets challenged with the combination of three pro-inflammatory cytokines (IL-1β, TNF-α and IFN-γ) reached up to 21 days with 75 % of cells surviving while less than 25% cells survived by day 4 in the absence of the anti-inflammatory peptide (Su et al., 2010). Furthermore, a synergetic protective effect had been observed for the IL-1RIP within the microgel, with even slightly enhanced insulin secretion in the presence of IL-1RIP in the microgel. Importantly, the hydrogel was produced by mixing 4-arm PEG macromers modified with cysteine and 4-arm PEG with thiol-end groups. The formed hydrogel network possessed the thiol groups that exhibited mild reductive properties and could protect encapsulated cells from oxidative stress (Robertson & Harmon, 2007; Pi et al., 2007).

Immobilization of cytokine antagonizing antibody within the hydrogel is complicated due to the large size of such a molecule (antibodies are usually of hundreds kDa), and protein stability and functioning can be altered during the immobilization process. Additionally, the presence of foreign antibodies in microgel structures can provoke host immune system response. Employing small molecules with the similar function can be one of the solutions such an issue. For example, small WP9QY peptide has been identified as a potential antagonist peptide that binds to human TNF-α. This peptide was used to produce the TNF-α-antagonizing PEG hydrogel (Lin et al., 2009). When the PEG hydrogel was formed via photoinitiated polymerization in presence of acryl-WP9QY and adhesive peptides in a precursor solution, encapsulated islets had higher insulin secretion even in the presence of TNF-α.

3.1.3 Cryopreservation of islets using microencapsulation

One of the common applications of microgel beads is cryopreservation of islets. Cryopreservation can be achieved through freezing to preserve the cells within the polymer matrix. However, during freezing islets can still be damaged by extracellular ice formation and cell shrinkage due to dehydration nature of freezing. Recently proposed vitrification approach allowed avoiding these issues. In this process, the presence of certain compounds, e.g., sugars (Doxastakis et al., 2005), stabilizes cells through a direct substitution of water molecules in the outer hydrophilic part of the membrane which leads to the formation of a glass that prevents fusion and reduces dehydration-induced stresses (Koster et al., 2000). Cryopreservation by freezing had been shown to have a significant negative effect on the structure of agarose gel microcapsules (Agudelo & Iwata, 2008). Cryopreservation changed the molecular network structure of agarose gel. Structural changes affected the diffusion of molecules within the gel. The small molecules, such as glucose, were still able to freely diffuse through porous gel, while, molecules of a medium size, such as insulin, had much lower diffusion coefficient after gel freezing. In contrast, vitrification process applied to the same materials had not introduced any changes in scaffolds. Islets preserved by vitrification do not suffer from ice formation and shrinkage and islets viability is drastically increased. Rat pancreatic islets were successfully cryopreserved by freezing in alginate beads (Schneider & Klein, 2011) or in poly(vinyl alcohol) macro-scaffolds (Qi et al., 2010). In the former case, islets were mixed with alginate solution of ultra-high viscosity prior to freezing. Beads then were formed using an air jet droplet generator to create a certain flow rate and finally cross-linked with barium ions. After 7 days of cryopreservation islets were able to maintain normal function *in vitro* and normalized non-fasting blood glucose in diabetic mice *in vivo*. However, since vitrification procedure requires the presence of low molecular weight molecules, some of them, such as poly(vinyl pyrrolidone) and PEG of a low molecular weight, can exert some toxicity to the vitrification solution. For example, insulin release ability of the islets vitrified in the presence of those compounds decreased by 45% after a vitrification/warming cycle (Agudelo & Iwata, 2008).

3.1.4 Encapsulation in epoxy-based microcontainers

New approach for islets microencapsulation has been introduced by Gimi et al. (Fig. 2) (Gimi et al., 2009). They proposed to use microcontainers made by polymerization of epoxy-based polymer. The microcontainer has a membrane containing nano-slots with the size of 25 nm that permit the bidirectional transport of metabolites and nutritional ingredients. Highly cross-linked epoxy-based polymer SU-8 with high glass transition temperature, thermal stability and high Young's modulus was used in fabrication of the microcontainer structure. The containers are capable of holding islets with dimensions upto $200 \times 200 \times 200 \mu m^3$. Overall device consists of two major components: a hollow cubical base and a nano-porous lid. The cubical base is intended for islets encapsulation, and the lid is sealed after an islet is placed inside the base. The base and the lid have a similar multilayer structure and a complicated forming procedure involving metal sputtering, polymer coating and nano-slot fabrication steps. Murine islets encapsulated into the microcontainers were viable after 48 hours in culture. The main advantage of this microcontainer approach is the possibility to directly visualize the encapsulated islets due to optical transparency of the microcontainer material. However, apart from complex fabrication, this approach might require additional steps for biocompatibility adjustment of the material and needs additional studies on islet long-term survival and functioning.

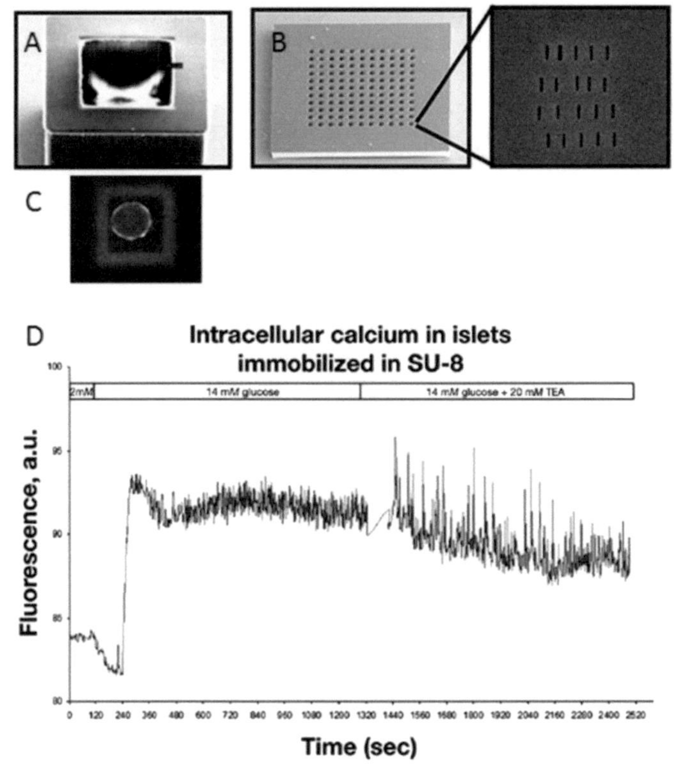

Fig. 2. Scanning electron microscope images of the microcontainer base (A) and lid (B) showing the 500-nm thin-membrane island structures. The thin membrane recessed within each island was milled with 25-nm slots using a focused ion beam. For encapsulated islets, the intracellular Ca^{2+} changes of an encapsulated islet in response to glucose and tetraethylammonium stimulations were measured using a spectral two-photon confocal microscope (C), and changes in the concentration of intracellular Ca^{2+} in response to glucose and tetraethylammonium were quantified as seen in this representative trace (D). Reprinted with permission from Gimi, et al., 2009 with permission from Journal of Diabetes Science and Technology.

3.2 Surface modification

Surface modification of living cells is a new approach in islet encapsulation. The main idea of surface modification is to reduce the size of implanted islets. This can be achieved by forming a conformal protective layer on the islet surfaces using various surface modification techniques. Both synthetic and natural polymers can be used to introduce various functional groups on the islets surfaces. Covalent binding to amino groups of cell membrane proteins, insertion of amphiphilic polymers into a cell membrane via hydrophobic interactions between a lipid bilayer of the cell membrane are general approaches employed in cell surface modification (Fig. 3). The covalent binding of protective layer to islet cell surfaces is expected to be stable and be present at the islet surface for certain periods of time.

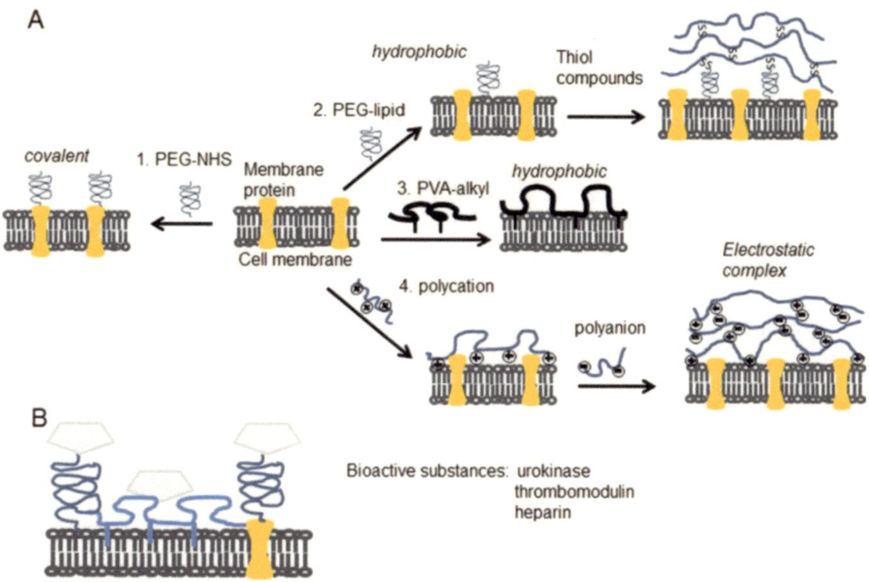

Fig. 3. (A) Modification of cell surfaces by covalent conjugation to cell surfaces (1), by incorporation of molecules into the cell membrane via hydrophobic interactions (2, 3), via electrostatic interactions of polyions with the cell membrane (4). Immobilization of bioactive molecules to cell surfaces via interactions of the cell membrane with polymers (B). Adapted from Teramura & Iwata, 2010 with permission from Elsevier.

3.2.1 Covalent conjugation of molecules to islet surfaces

Thin conformal coatings based on PEG are highly attractive for cell surface modification due to its biocompatibility and protein-resistant properties based on PEG low interfacial free energy with water, its unique properties in aqueous solutions, its high surface mobility, and its substantial steric stabilization effects (Amiji & Park, 1993). PEGylation has been used for camouflaging islet surfaces to immunologically protect transplanted islets from the immune system (Panza et al., 2000; Lee et al., 2002; Lee et al., 2004). Methoxy-PEG-succinimidyl propionic acid molecules conjugated with amino groups of collagen matrix at the islet surface through a stable amide bond protected islets from immune cell attack (Fig. 3, A1). The conjugated PEG molecules have been shown to block immune cells from diffusing to the transplanted islets, which allowed islets to function in the diabetic recipients for several weeks (Lee et al., 2006a; Lee et al., 2006b; Lee et al., 2007). Another advantage of the nano-thin PEG layer is the small size of the produced shell compared to the size of encapsulated islets. The small shell size and the demonstrated protection against the host immune system should allow transplantation into the portal vein by catheter injection and still prevent immune response of the host.

To achieve grafting of the end-functionalized PEG polymer onto the islet surfaces, islets are cultured in media in the presence of various concentrations of the polymer. To increase the amount of a grafted polymer on surfaces, the incubation time can be increased or the incubation step can be repeated. The grafting of methoxy-PEG (mPEG) onto the islets surface is usually carried out via the succinimidyl ester end groups introduced in mPEG. These groups couple to amino groups present in the collagen matrix around islets. Controlling the grafting time is important since longer grafting times can lead to diffusion of PEG inside the islets which may increase a chance of swelling and exposure of islets to outside environment. Furthermore, long grafting time increases the probability of islet damaging (Lee et al., 2002).

The molecular weight of the grafted PEG is another important parameter for islet surface modification. Barani et al., showed that grafted mPEG of 5 and 10 kDa onto islets isolated from Wistar rats had a different effect on cell functioning and viability (Barani et al., 2010). Insulin secretion for the islets grafted with 5 kDa mPEG was at the same level as for unmodified islets, while overall insulin secretion from islets modified with 10 kDa mPEG decreased. However, its protective ability was higher which correlates well with the decrease in mesh size and thus less exposure to immune system. The effect of PEG molecular weight was also studied on porcine islets (Xie et al., 2005). The highest protection ability without affecting islets functional capability was found for 5-6 kDa PEG that allowed complete coverage of islets surface.

In the case of two reactive succinimidyl ester end groups, additional functional molecules such as albumin can be brought to the islet surfaces (Xie et al., 2005). Introducing albumin and PEG on the surface can be used to suppress the immunogenic reactions (Hortin et al., 1997). Presence of human albumin not only increased cytoprotection of islets but also significantly increased the insulin production which was possible due to increasing density of protective coverage with albumin presence. The so-modified islets maintained their functionality *in vivo* up to 15 days.

Heparin is a highly sulfated glycosaminoglycan and is very often used as an injectable anticoagulant. Systemic delivery of heparin at therapeutic doses, however, substantially increases the risk of bleeding. Furthermore, the effect of soluble heparin is limited to 2-3 hours *in vivo* and has no significant effect on long-term IBMIR reaction (Bennet et al., 1999). At the same time, heparin, immobilized on artificial surfaces which mimic the protective ability of the endothelial cells, demonstrated inhibition of coagulation and complement activation (Bennet et al., 1999). Immobilized heparin coating was successfully made on human, porcine and mouse islets by step-by-step incubation procedure (Cabric et al., 2007). Islet surfaces were first biotinylated through covalent attachment of succinimidyl ester-biotin, then incubated with avidin and finally, heparin conjugate was covalently attached on the modified islets surface (Fig. 3, B). The heparin coating produced by this method was present of the islets surfaces at least 72 hours, though not detectable after 4-5 weeks of transplantation.

Islets surface modification with other inhibitors of the coagulation cascade through bifunctional PEG linkers can provide another strategy to reduce IBMIR. Human recombinant thrombomodulin was conjugated to the islet surface through bifunctional PEG grafts (Stabler et al., 2007). Thrombomodulin is an endothelial cell transmembrane protein which can deactivate procoagulant–proinflammatory properties of thrombin through its binding (Esmon et al., 2004). The formed thrombomodulin–thrombin complex leads to rapid inactivation of clotting factors Va and VIIIa and reduction of new thrombin generation (Esmon et al., 1982). This approach of thrombomodulin conjugation allowed for the substantial delay of clot formation upon incubation of islets in human plasma with preserved islet viability and glucose-stimulated insulin secretion capability.

Despite the immunoprotective capabilities, the camouflaging layer of PEG cannot provide a long-term protection against small cytotoxic molecules produced by the immune system in response to metabolic activity of encapsulated islets after the transplantation. A combination of a protective coating and a low dose of an immunosuppressive drug can be a way to provide a suppression of host immune system without causing highly toxic effects. A low dose of Cyclosporine A has been shown to effectively prevent rejection of PEG-modified islets and support their functioning and survival up to 1 year (Lee at al., 2007). Similar effect was demonstrated by employing 6-arm-PEG-catechol that allowed for a coating with a higher density and a lower amount of immunosuppressive drug Tacrolimus (Jeong et al., 2011). Catechol is a surface independent anchor molecule which ensured conjugation of PEG-catechol to collagen matrix around the islet. The survival time of 6-arm-PEG-catechol grafted islets was similar to that of unmodified islets. However, the administration of the drug increased the survival time of 6-arm-PEG-catechol grafted islet almost two times.

3.2.2 Incorporation of molecules in the cell membrane

Islets surface modification can be achieved via hydrophobic interactions of amphiphilic polymers, such as PEG-phospholipids and PVA with long alkyl chains, with lipids in cell membranes (Takemoto et al., 2011; Chen et al., 2011). Presence of hydrophobic parts, phospholipids and alkyl chains, in polymer and cell membranes are responsible for spontaneous incorporation of co-polymers into the lipid layers of cell membrane. The spontaneous incorporation is greatly affected by the length and hydrophobicity of alkyl chains (Rabuka et al., 2008; Teramura et al., 2007). Using this approach, the PEG-phospholipid-based coating was modified with fibrinolytical enzyme urokinase and with a soluble domain of the anticoagulant, thrombomodulin (Fig. 3, B) (Chen et al., 2011). The immobilization of heparin, urokinase, or thrombomodulin can improve the graft survival after the transplantation. The maleimide-PEG-lipids were utilized to immobilize proteins on the islets surface through reaction of maleimide and thiol-modified proteins. However, thiol groups can easily form disulfide bonds under physiological conditions and that decreased the efficiency of the conjugation.

Urokinase is a serine protease that activates plasminogen into plasmin, which dissolves fibrin blood clots. Urokinase can be conjugated onto the surface of islets to dissolve blood clots surrounding islets in the liver for inhibition of the cascade reactions (Takemoto et al., 2011). Hybridization between urokinase and DNA-PEG-lipids was made due to complementary sequences on the protein and single strain DNA (Fig. 4). DNA hybridization method is versatile for conjugation of various bioactive molecules. Using different sequences, it was possible to conjugate different molecules on the cell surfaces. The cell morphology was not affected by such modification, although, urokinase activity disappeared after 4 days in culture.

Overall, surface modification achieved via hydrophobic interaction and insertion of hydrophobic chains into the cell membrane demonstrated a limited retention time. For PEG-lipids this time depends on the alkyl chain length of lipid chains (Inui et al., 2010). The longer alkyl chains allowed PEG-lipid molecules to retain longer on the cell membrane. However, the overall retention time was no longer than 48 hours before the all molecules disappeared from the islets surfaces. Incorporation of hydrophobic molecules into the cell membrane leads to their uptake into the cell cytoplasm. Adding amphiphilic part to the structures prevents the uptake but molecules are released into the surrounding media.

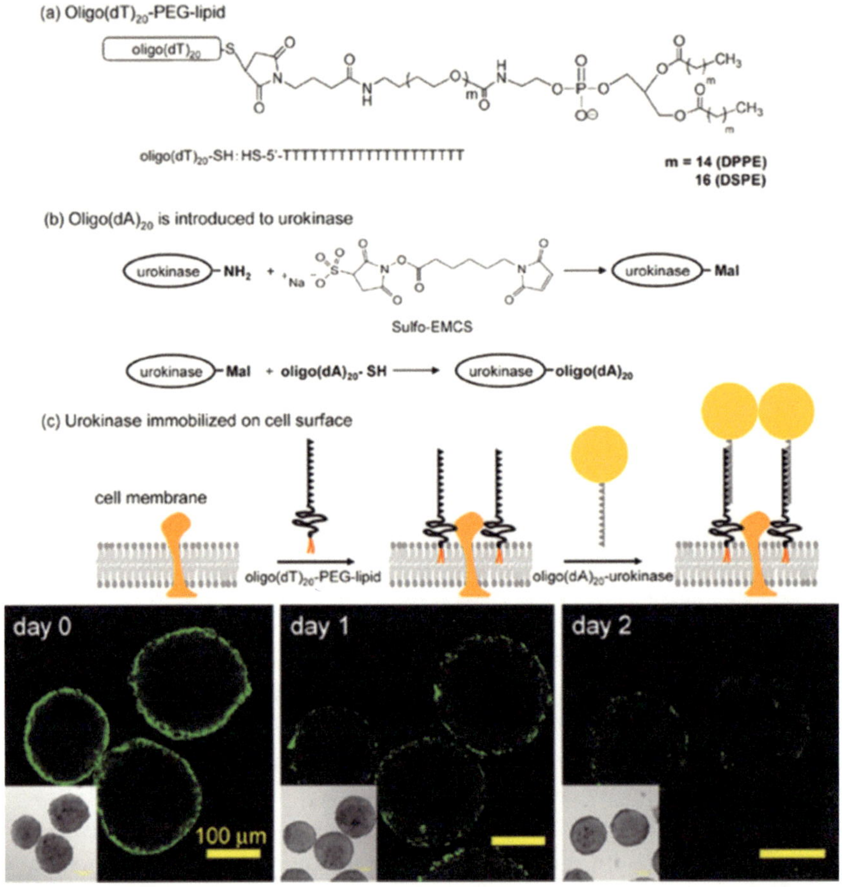

Fig. 4. Top: (a) Chemical structure of oligo(dT)20-conjugated PEG-lipid (oligo(dT)20-PEG-lipid). (b) Urokinase is modified with Sulfo-EMCS to introduce maleimide groups, and then oligo(dA)20-SH is conjugated on urokinase through the thiol/maleimide reaction.
(c) Oligo(dT)20-PEG-lipid is incorporated into the cell surface by the hydrophobic interaction between alkyl chains and the lipid bilayer of the cell membrane. Oligo(dA)20-urokinase is applied to cells carrying oligo(dT)20. Urokinase is conjugated on the cell surface through oligo(dT)20 -oligo(dA)20 hybridization. Bottom: Confocal laser scanning microscopic images of urokinase--modified-islets which were subjected to immunostaining for urokinase. Islets were modified with oligo(dT)20-PEG-lipids and then oligo(dA)20-urokinase. The urokinase-modified-islets just after treatment, after 1 and 2 days of culture. Reprinted with permission from Takemoto et al., 2011. Copyright 2011 American Chemical Society.

3.2.3 Surface modification through electrostatic interactions

Heparin has an affinity to a number of plasma proteins and growth factors including endothelium and vascular endothelium growth factors. Employing this property of heparin, the synthetic heparin-binding peptide amphiphile was used to immobilize heparin and, subsequently, growth factors on islet surfaces (Chow et al., 2010). Initial immobilization of the amphiphile on the islet surfaces can be achieved *via* electrostatic interactions of positively charged amphiphile and negatively charge cell membrane (Fig. 3, A4). Later, negatively charged heparin can be bound to the amphiphile. The so-formed amphiphile-heparin complex coating on the islets surface was able to bind and retain growth factors up to 48 hours. The complex nanostructures provide extracellular matrix-like scaffold for the encapsulated islets for supporting their viability. Such immobilization of growth factors does not affect viability or function, yet, they are able to stimulate an angiogenic response in the islets.

3.2.4 Immobilization of living cells on islet surfaces

Endothelial cells on the islets surfaces have a good tolerance towards blood presence and can provide a protection against IBMIR. Introduction of such protective cells around islets can be used as an effective protection strategy. The human aortic endothelial cells were introduced onto isolated human islets of Langerhans by mixing of both cell types and incubation for several hours (Johansson et al., 2005). The clotting can be significantly reduced with the 90%-cell-coating present on islet surfaces. The consistency of the cell amounts in such coating can be improved using the PEG-phospholipid-based approach (Teramura & Iwata, 2009; Teramura & Iwata, 2008a). In that case, islets and human endoderm kidney cells were first separately biotinylated through biotin-PEG-lipid anchors to cell membranes with further streptavidin coating on the kidney cells (Fig. 5). Immobilization of the endothelium cells around islets was then made via streptavidin-biotin reaction. The cell enclosure was stable on the islet surfaces within 3-5 days *in vitro*. To overcome the streptavidin immunogenicity, the cell deposition was also made via the PolyDNA-PEG-lipid conjugate (Teramura & Chen, 2010; Teramura & Minch, 2010). Human endoderm kidney cells were able to rapidly proliferate forming a cell multilayer on the islets surfaces protecting the encapsulated islets from the host immune response. However, the cell oxygen consumption can result in lowered oxygen available for the encased islets. Thus, additional studies are necessary to clarify the short- and long-term effects of the cell presence on islets surfaces.

3.3 Conformal coating of islets

3.3.1 Layer-by-layer (LbL) approach

The layer-by-layer (LBL) assembly of polymers based on sequential adsorption of oppositely charged components is one of the established methods for the preparation of thin polyelectrolyte multilayer films with controlled properties. The LBL represents a universal surface modification approach that allows for producing surface-attached films with controlled thickness, permeability, mechanical properties and surface chemistry. The technique has been recently applied to modify islet surfaces (Krol et al., 2006; Wilson et al., 2008). The LbL modification of islet surfaces is based on alternating deposition of water

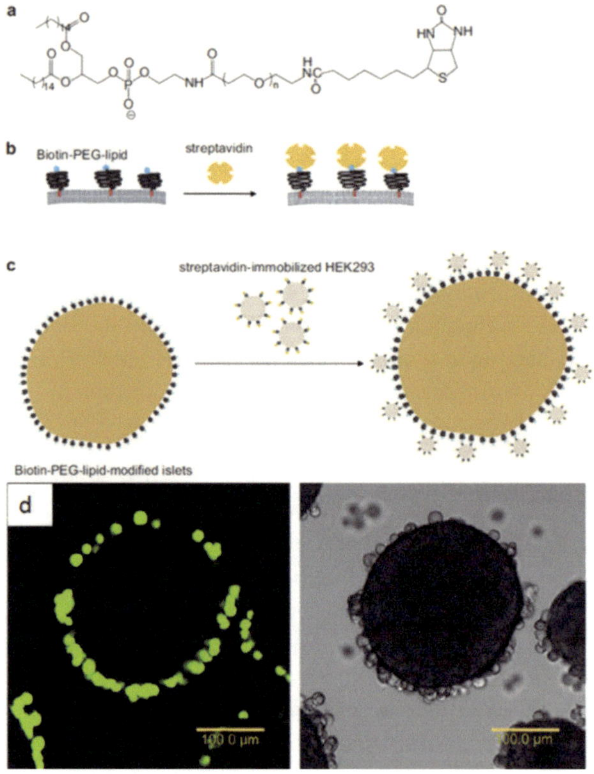

Fig. 5. (a) Chemical structure of biotin–PEG-conjugate (biotin–PEG–lipid). (b) Schematic illustration of the interaction between streptavidin and biotin–PEG–lipid at the lipid bilayer cell membrane. Biotin–PEG–lipid has hydrophobic acyl chains and is incorporated into the cell surface by anchoring into the lipid bilayer. Streptavidin is immobilized on the cell surface by anchoring to biotin–PEG–lipid. (c) Scheme for the immobilization of streptavidin-immobilized HEK293 cells on the surface of biotin–PEG–lipid-modified islets. After mixing streptavidin-immobilized HEK293 cells and biotin–PEG–lipid-modified islets, they were cultured in medium at 37°C on a culture dish. During culture, HEK293 cells were spread and grown on the cell surface to cover the whole surface. (d) Hamster islets modified with biotin–PEG–lipid and immobilized with streptavidin-immobilized HEK293 cells. The HEK293 cells were labeled with CellTracker. Reprinted from Teramura & Iwata, 2009 with permission from Elsevier.

soluble polymers on surfaces from aqueous solutions which results in nano-thin coatings of controllable thickness and composition (Decher & Schlenoff, 2002; Kharlampieva & Sukhishvili, 2006; Tang et al., 2006). The ultrathin conformal coating affords a faster response to stimulation and the possibility to bind factors or protective molecules to the protective ultrathin shell with the later slow triggered release of these molecules (Chluba et al., 2001). By selecting specific polyelectrolytes, a defined cutoff of the coating (Kozlovskaya & Sukhishvili, 2006) is possible, as is inhibitor binding to prevent graft rejection, microphage attacks, or antibody recognition (Kim & Park, 2006). Modification of the last coating layer can be used to support functionality of islets and reduce the immune response from a host system. The cutoff of the polyelectrolyte multilayer (PEM) is defined by polyelectrolytes used in coating formation (Krol et al., 2006).

3.3.2 The ionic LbL assembly

To promote a multilayer film formation on the cell surfaces, the negatively charged cell surface is treated with a cationic polymer solution and the cell surface is further exposed to an anionic polymer solution to form an electrostatically-paired polyelectrolyte complex film (Fig. 3, A4). Effect of molecular weight of polyelectrolytes and the charge of outermost layer was demonstrated in case of the LBL encapsulation of human islets into poly(allylamine hydrochloride)/poly(styrenesulfonate sodium salt) (PAH)/(PSS) and poly-(diallyldimethylammonium chloride) (PDADMAC)/PSS layers. Islets encapsulated into PAH/PSS and PDADMAC/PSS multilayers using a higher polycation molecular weight demonstrated a limited insulin release due to a lowered permeability of insulin through the polyelectrolyte membrane. A decrease in a polycation molecular weight resulted in larger pores of the polyelectrolyte membrane and restored responsive relationship between glucose stimulation and insulin response of the coated islets (Krol et al., 2006).

Most cationic polymers widely used in the LbL modification of surfaces such as poly(L-lysine) (PLL) and poly(ethylene imine) (PEI) are extremely cytotoxic and cells treated with the polycations can be severely damaged. Their cytotoxic effect though has been observed to be dependent on polycation concentration and exposure time (De Koker et al., 2007). The overall cytotoxicity of the polyelectrolytes originates from positive charge of polycations which can induce pore formation within the cell membrane causing its damage and, eventually, cell death (Bieber et al., 2002; Godbey et al., 1999). The high toxicity of the PAH/PSS LbL film was confirmed by Wilson et al (Wilson et al., 2008). They demonstrated that coating the murine islets with only 3 layers of PAH/PSS/PAH led to the reduction of islet viability by 70%. Similar effect was found for islets coated with 3 layers of PLL/alginate LbL film. Even 15 minutes of islets incubation with low concentration of PLL results in ~60% decrease in cell viability. Menger et al showed that PLL was able to pass through the lipid bilayer if it was previously allowed to form complex with anionic lipids (Menger et al., 2003). PEI was found extremely toxic to the islets. This polycation destroys the cell membrane immediately after its interactions with the membrane surface (Teramura et al., 2008b). The overall charge arrangement of a polycation and its interaction with the cell membrane strongly depends on the three-dimensional structure and flexibility of the polymer chains. It has been shown that polymers with highly flexible chains and a high cationic density will exert tremendous cytotoxicity. Thus, the polycations with globular structures demonstrated good biocompatibility, whereas polymers with more linear and flexible structure such as PLL and PEI showed higher cytotoxicity (Teramura et al., 2008b).

Since the polycations toxicity partially depends on the polymer charge density, it can be attenuated by conjugating neutral molecules, such as PEG, to the critical number of amino groups along the polycation backbones. PEGylation of PLL is carried out through grafting of N-hydroxysuccinimide-PEG (NHS-PEG) chains to amino groups on PLL backbone to produce PLL-g-PEG. The grafted PEGs are unbranched, hydrophilic, discrete-length molecules in the form of Methyl-PEGn-NHS ester, where the subscript "n" denotes a number of the ethylene glycol units. The NHS ester end group is spontaneously reactive with primary amines, providing for efficient PEGylation of amine-containing molecules or surfaces. The methoxy(ethylene glycol) grafts were conjugated to PLL backbone through a covalent attachment to lysine residues (Wilson et al., 2009). Forty percent of PEG substitutes on the PLL chain allowed for attenuation of the PLL positive charges without any

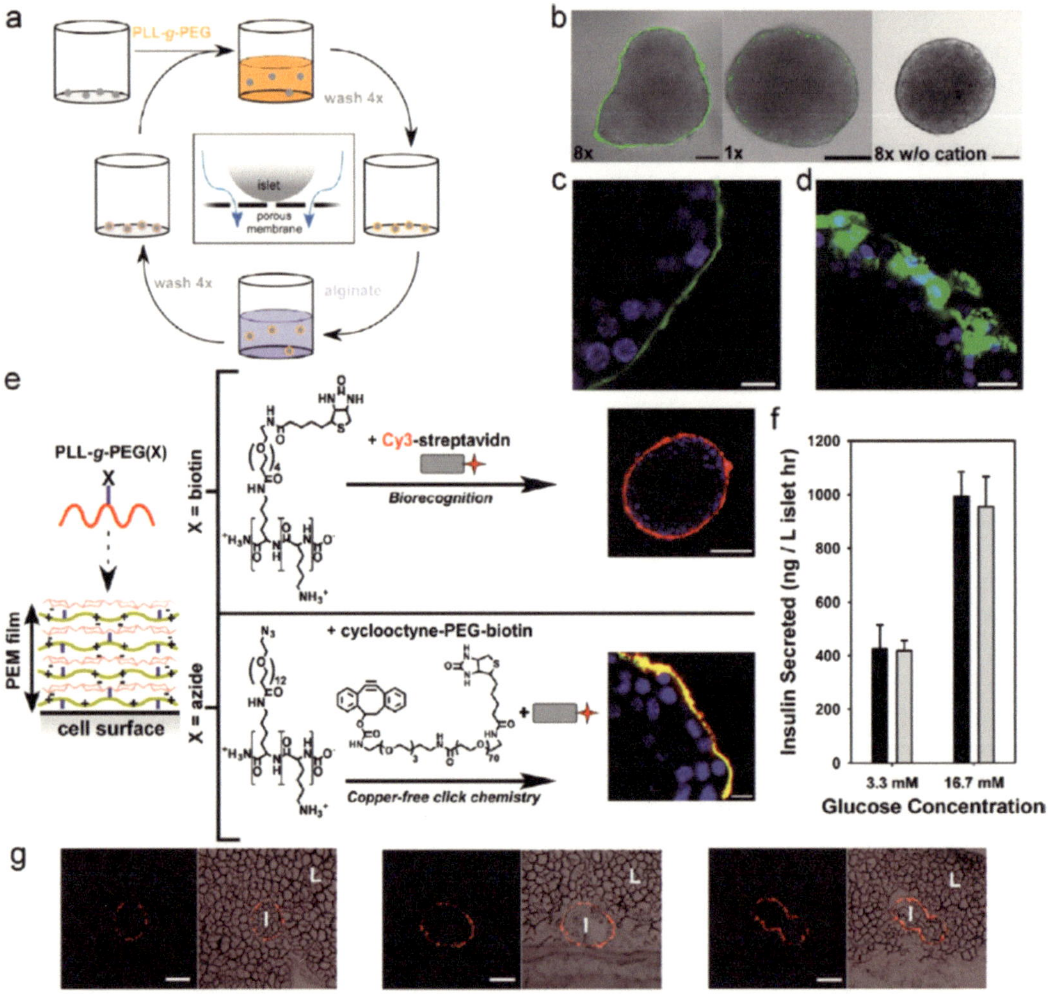

Fig. 6. Cell-surface-supported PEMs were assembled on individual pancreatic islets through LbL deposition of PLL-g-PEG copolymers and alginate. (a) Method to assemble PEMs on islets. (b) Representative confocal micrographs overlaid on bright-field images of coated islets coated using flourescein-labeled alginate (F-Alg) with eight bilayers (8x), a single bilayer (1x), or treated only with F-Alg (8x, w/o cation). (c) F-Alg is localized on the extracellular surface of cells, confirming the cell-surface-supported nature of films. (d) Deposition of a single PLL/F-Alg bilayer resulted in intracellular internalization of alginate by peripheral cells. (e) Chemistry and reactivity of cell-surface-supported films can be tailored through integration of biotin- and azide-functionalized PLL-g-PEG copolymers. (f) Insulin secretion by islets coated with a (PLL-g-PEG /alginate)$_8$ film (gray) and untreated islets (black) in response to a step-change in glucose. (g) Confocal (left) overlaid on bright-field micrographs (right) of frozen sections of liver (L) after intraportal transplantation of islets (I) engineered with PEM films labeled with streptavidinCy3. Scale bars: b,e (top), g=50 µm; c,d,e (bottom)=10 µm. Reprinted with permission from Wilson et al., 2011. Copyright 2011 American Chemical Society.

deleterious effect on the islet viability (Fig. 6). The modification of PEG grafts with various functional molecules, such as biotin, hydrazide and azide, can extend the functional capabilities of the PLL-g-PEG-based LbL islet coatings. For example, the deposition of the PLL-g-PEG-biotin outmost layer on top of the PLL-g-PEG/Alginate multilayer film can generate surface densities of biotin functional groups comparable with that obtained by the treatment of islet surfaces with only NHS-PEG-biotin molecules (PEGylation/biotinylation through the NHS-ester coupling). Unlike the latter, the former approach is advantageous as it does not alter cell surface morphology and allows for controlled densities of the biotin on the modified islet surfaces (Krishnamurthy et al., 2010). The first successful *in vivo* transplantation of PEM engineered islets was demonstrated for the murine islets coated with conformal PLL-g-PEG/Alginate LbL films (Fig. 6g) (Wilson et al., 2011).

High toxicity of most polyelectrolyte polycations limits their use in biomedical applications. However, natural biopolymers chitosan and alginate have more similarities with the extracellular matrix, are chemically versatile and have a good biocompatibility. These linear polysaccharides carry opposite charge and can be electrostatically bound in a PEM. The $(Chitosan/Alginate)_3$ islet coating was achieved *via* alternate deposition starting from positively charged chitosan (Zhi et al., 2010). The deposition conditions had been shown to greatly influence the islets viability, which well correlates with difference in charge density and toxicity of the polycation at high and neutral pH values. Additional protective outermost coating layer of phosphorylcholine (PC)-modified chondroitin-4-sulfate was introduced to reduce non-specific protein adsorption of the film. PC-moieties demonstrated remarkable repelling and hemocompatible properties. Increase in the coating thickness by adding additional layers of (Chitosan/Alginate) up to 5 bilayers did not adversely affect islets viability and insulin release, and the coated islets were viable up to 5 weeks of post-encapsulation.

3.3.3 The LbL assembly based on covalent and/or specific interactions

A multilayer PVA membrane formed via the layer-by-layer assembly was investigated for islet immunoprotective capabilities (Fig. 3, A3) (Teramura et al., 2007). In this approach maleimide-PEG-conjugated phospholipids were used to first modify the cell membrane surface to promote further interactions with PVA derivatives. It is well-known that PEG-conjugated phospholipids can be immobilized on the cell membrane through incorporation of the lipid chains into the cell membrane due to their hydrophobic interactions with the lipid bilayers of the membrane (Iwata et al., 1992). Moreover, PEG-phospholipids are more compatible with cells compared to polycations used in the ionic LbL surface modification of islets. A layer of PVA with introduced thiol groups (PVA-SH) was covalently attached to the maleimide-PEG anchors via thiol-maleimide reaction. The LbL multilayer of PVA was then deposited by alternating immersion of the islets into PVA-SH and PVA-pyridyl disulfide (PVA-PD). The driving force for the multilayer formation was a thiol/disulfide exchange reaction between the PVA derivatives. This ultra-thin PVA membrane affected neither cell viability nor insulin release function.

The LbL assembly of heparin multilayers was investigated for suppression of instant blood-mediated inflammatory reactions for the case when islets are to be transplanted through the portal vein to liver (Luan et al., 2011). The heparin well-known for its anti-thrombogenic properties was co-assembled with human soluble form complement receptor 1 (sCR1) which is a potent inhibitor of the classical and alternative complement activation pathways (Pratt et al., 1996). The islets surfaces were first modified with PEG-phospholipid conjugates bearing maleimide groups (Fig. 7). The thiol-modified sCR1 molecules were then covalently bound to islets through the maleimide-PEG-lipid anchors via thiol-maleimide reaction. Following heparin immobilization with sCR1 was possible due to its strong affinity to sCR1. It is known though that covalent immobilization does not allow for long-term stability of the immobilized molecules at the cell surfaces due to fast degradation of the conjugation bonds. For instance, PEG-lipids gradually disappeared from the cell surface without uptake into the inside of cells dissociating from the cell surface into the medium (Teramura et al., 2008b). However, increase in the number of heparin /sCR1 layers upto 3 bilayers allowed subsequently increase the heparin and sCR1 retention time on islets surfaces up to 8 days *in vitro* with the gradual release from the islet surfaces over longer time periods. As the severe IBMIR reactions, especially the activation of the coagulation system and the complement cascade, occur in the first 2 days after transplantation, the heparin/sCR1 LBL approach can be a promising method for protecting the transplanted islets. Nevertheless, the protective effects and anti-thrombin activity of the LbL coatings are to be yet evaluated *in vivo.*

The LbL coating of islets based on streptavidin/biotin anchoring as a driving force for the LBL was also developed (Dai et al., 2007; Kizilel et al., 2010). PEGylated multilayers were grown using LbL assembly of biotinylated PLL and streptavidin. PEG was incorporated into the LbL architectures by assembly with biotin-derivatized PLL-g-PEG (Dai et al., 2007). In another example, the rat islet cell membranes were first biotinylated through covalent reaction of the membrane proteins and NHS-ester groups of NHS-PEG-biotin. The second layer of streptavidin was followed by a layer of a biotin-PEG conjugate with the insulinotropic Glucagon-like Ppetide 1 (GLP-1) (Kizilel et al., 2010). GLP-1 is a 30-amino acid peptide hormone produced in the intestinal epithelial endocrine L-cells. The main action of GLP-1 is to stimulate insulin secretion, acting as an incretin hormone, and to stimulate islet growth and neogenesis. Insulinotropic activity of GLP-1 is glucose-dependent and exerted via interaction with the GLP-1 receptor located on the cell membrane of the β-cells. Stimulation of insulin secretion due to GLP-1 immobilization at the islet surfaces can result in reducing the number of islets necessary for normalization of blood glucose of a patient. GLP1-PEG-Biotin was immobilized as an outermost layer to enhance insulin secretion. Islets coated with PEG-GLP-1 demonstrated enhanced insulin secretion and no time-delay in response to high glucose level. Streptavidin, however, is an immunogenic protein derived from bacteria *Streptomyces avidinii* and in most cases is not suitable for use in clinical studies.

Obviously, multilayer PEGylation technique provides an advantage over the single layer via enhancing immune-protective barrier but, on the other hand, the LbL method based on subsequent immersion of islets in alternating polymer solutions can be time-consuming and thus inappropriate for clinical use. The ultra-thin multilayer build-up process was further modified by Hume et al to overcome this issue (Hume et al., 2011). In their approach, the hydrogel was formed via photoinitiated polymerization of PEG diacrylate (PEGDA) or methacrylate pre-cursor solutions. Proteins such as immunoglobulins, and intercellular adhesion molecule-2 (ICAM-2) were immobilized on the coating surface via thiol-(meth)acrylate co-polymerization. The methacrylate-PEG was found to be more cytocompatible compared to acrylate-PEG and demonstrated protective effects against T cells.

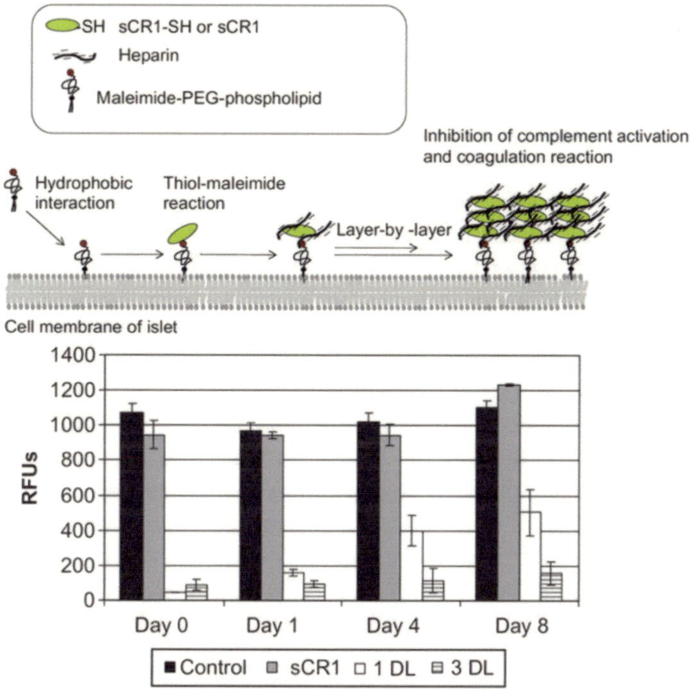

Fig. 7. Immobilization of sCR1and heparin on the islet cell surface (top). Relative thrombin inactivation activities of multiple sCR1-heparin layers (1 double-layer (1DL), three-double layer (3 DL) on glass plates. The activities were determined after the glass plates were maintained in culture medium (medium RPMI-1640 containing 10% FBS) for the indicated periods (bottom). Reprinted from Luan et al., 2011 with permission from Elsevier.

Thus, despite the advantages the ultrathin coating can present, chemical modification of the cell membrane using covalent conjugation of molecules, for example, covalent PEG binding via NHS-ester conjugation to the cell membrane proteins, can result in damaging of the membrane proteins and, consequently, in cell physiology disturbance. Moreover, the immunoisolation capabilities of ultra-thin hydrogel-based multilayers are still limited because of poor control over selective permeability towards small cytotoxic cytokines, and hypoxia within the gel environments. Current investigations are therefore focused on synthesis of hydrogels containing bioactive moieties, rather than unmodified coatings, aiming at "smart" protection of encapsulated islets from the damage originating from T-cells and small free radical species.

4. Outlook and perspective

Though pancreatic islet transplantation has emerged as a promising treatment for diabetes, its clinical application, however, remains limited due to serious side effects of immunosuppressive therapy necessary to prevent host rejection of transplanted islets.

Lifelong requirement of immunosuppressive drugs has deleterious effects on β-cell function and on host's ability to fight disease. To protect islets from immune-mediated destruction, camouflaging the islet surfaces is necessary for immunoisolation and immunoprotection. Current islet modification strategies are challenging for transplantation due to interference with islet functioning, limited nutrient transport, and/or cytotoxicity. Thus, current strategies mostly involve either imbedding islets within thick hydrogels that have limited nutrient transport and require large injection volumes or covalent conjugation to islet surfaces, which can interfere with cell function.

The ultra-thin coating approach possesses several advantages. Islets modified with such coatings can be easily implanted into liver through the portal vein, the preferable site for islet implantation. Islet necrosis can be avoided because of the rapid diffusion of nutrients and oxygen through the coating. Moreover, due to a minimal volume of the enclosed islets, an adequate release of insulin can be achieved in response to blood glucose changes. Though, despite the advantages, chemical modification of the cell membrane using covalent conjugation of molecules to the cell membrane proteins can result in cell physiology disturbance. In this respect, non-covalent modification method based on LbL can afford for cytocompatible coatings if non-toxic components are used. The ultrathin LbL coating allows for faster response to stimulation and the possibility to bind factors or protective molecules to the protective ultrathin shell with the later slow triggered release of these molecules. By selecting specific polyelectrolytes, a defined cutoff of the coating is possible, as is inhibitor binding to prevent graft rejection, microphage attacks, or antibody recognition. Another important point is that the mechanical properties of the LbL-based films can be adjusted through an introduction of appropriate components or the chemical modification of the constituents. This issue is particularly important as β-cell in mature islets can still grow (Dor et al., 2004) causing the islet size increase. Unlike hydrogel-based ultrathin coatings, LbL coatings have the ability to withstand islet size increase and effectively enclose the islets for long periods of time. Another crucial issue is the preservation of the islet integrity because functionality losses were observed when islets disintegrated or fused in suspension culture. In this respect, the LbL technique provides a conformal and stable coating of individual islets.

However, despite the significant promise of the LbL strategy for islet modification, the main drawback of the approach is cytotoxicity of the used compounds. Moreover, ultra-thin encapsulating films may not work as reliable barriers against free-radicals. Indeed, thick agarose microbeads or alginate microcapsules can provide a more effective shield capable of inactivating free radicals. This issue is particularly crucial since most inflammatory processes are associated with oxidative stress initiated by production of reactive oxygen species which can function as signaling molecules in many cell types.

Considering the previous studies, it is important to develop new strategies for design of new multifunctional coatings with immunomodulatory capabilities which can permit the reestablishment of ECM support and maintain the physiological needs of the islets. In this respect, the LbL approach offers opportunities for integration of the inherent advantages of both islet microencapsulation and surface modification approaches. For example, multifunctional LbL materials designed from non-toxic biologically-active polymers can provide novel immunoprotective and anti-inflammatory coatings crucial for prolonged islet viability and functions.

5. Acknowledgment

The work was supported by Award #P30EB011319 from the National Institute of Biomedical Imaging and Bioengineering at NIH.

6. References

Abalovich, A., Jatimliansky, C., Diegex, E., Arias, M., Altamirano, A., Amorena, C., Martinez, B., & Nacucchio, M. (2001). Pancreatic islets microencapsulation with polylactide-co-glycolide. *Transplant. Proc.*, Vol.33, No.1-2, pp.1977–1979, ISSN 0041-1345

Agudelo, C.A., & Iwata, H. (2008).The development of alternative vitrification solutions for microencapsulated islets. *Biomaterials*, Vol.29, No.9, pp.1167-1176, ISSN 0142-9612

Amiji, M., & Park, K. (1993). Surface modification of polymeric biomaterials with poly(ethylene oxide), albumin, and heparin for reduced thrombogenicity. *Journal of Biomaterials Science - Polymer Edition.*, Vol.4, No.3, pp. 217-234, ISSN 0920-5063

Barani, L., Vasheghani-Farahani, E., Lazarjani, H.A., Hashemi-Najafabadi, S., & Atyabi, F. (2010). Effect of molecular mass of methoxypoly(ethylene glycol) activated with succinimidyl carbonate on camouflaging pancreatic islets. *Biotechnol. Appl. Biochem.*, Vol.57, No.1, pp.25-30, ISSN 0885-4513

Basta, G., Sarchielli, P., Luca, G., Racanicchi, L., Nastruzzi, C., Guido, L., Mancuso, F., Macchiarulo, G., Calabrese, G., Brunetti, P., & Calafiore, R. (2004). Optimized parameters for microencapsulation of pancreatic islet cells: an in vitro study clueing on islet graft immunoprotection in type 1 diabetes mellitus. *Transpl. Immunol.*, Vol.13, No.4, pp.289-296, ISSN 0966-3274

Beattie, G. M., Montgomery, A.M.P., Lopez, A.D., Hao, E., Perez, B., Just, M.L., Lakey, J.R.T., Hart, M.E., & Hayek, A. (2002). A novel approach to increase human islet cell mass while preserving beta-cell function. *Diabetes*, Vol.51, No.12, pp.3435-3439, ISSN 0012-1797

Beck, J., Angus, R., Madsen, B., Britt, D., Vernon, B., & Nguyen, K.T. (2007). Islet Encapsulation: Strategies to enhance islet cell functions. *Tissue Eng.*, Vol.13, No.3, pp.589-599, ISSN 1076-3279

Bennet, W., Sundberg, B., Groth, C.G., Brendel, M.D., Brandhorst, D., Brandhorst, H., Bretzel, R.G., Elgue, G., Larsson, R., Nilsson, B., & Korsgren, O. (1999). Incompatibility between human blood and isolated islets of Langerhans: a finding with implications for clinical intraportal islet transplantation? *Diabetes*, Vol.48, No.10, pp.1907-1914, ISSN 0012-1797

Berg, M.C., Yang, S.Y., Hammond, P.T., Rubner, M.F. (2004). Controlling mammalian cell interactions on patterned polyelectrolyte multilayer surfaces. *Langmuir*, Vol.20, No.4, pp.1362–1368, ISSN 0743-7463

Bieber, T., Meissner, W., Kostin, S., Niemann, A., & Elsasser, H.P. (2002). Intracellular route and transcriptional competence of polyethylenimine-DNA complexes. *J. Control. Release*, Vol.82, No.2-3, pp.441-454, ISSN 0168-3659

Bünger C.M., Gerlach, C., Freier, T., Schmitz, K.P., Pilz, M., Werner, C., Jonas, L., Schareck, W., Hopt, U.T., & De Vos, P. (2003). Biocompatibility and surface structure of chemically modified immunoisolating alginate-PLL capsules. *J. Biomed. Mater. Res. A*, Vol.67, No.4, pp.1219-1227, ISSN 1549-3296

Cabric, S., Sanchez, J., Lundgren, T., Foss, A., Felldin, M., Källen, R., Salmela, K., Tibell, A., Tufveson, G., Larsson, R., Korsgren, O., & Nilsson, B. (2007). Islet surface heparinization prevents the instant blood-mediated inflammatory reaction in islet transplantation. *Diabetes*, Vol.56, No.8, pp.2008-2015, ISSN 0012-1797

Calafiore, R., Basta, G., Luca, G., Boselli, C., Bufalari, A., Giustozzi, G.M., Moggi, L., Brunetti, P. (2006). Alginate/polyaminoacidic coherent microcapsules for pancreatic islet graft immunoisolation in diabetic recipients. *Ann N.Y. Academy Sci.*, Vol.831, pp.313-322, ISSN 0077-8923

Chandy, T., Mooradian, D.L., & Rao, G.H.R. (1999). Evaluation of modified alginate-chitosan-poly(ethylene glycol) microcapsules for cell encapsulation. *Artif. Organs*,Vol.23, No.10, pp.894–903, ISSN 0160-564X

Chen, J.P., Chu, I.M., Shiao, M.Y., Hsu, B.R.S., & Fu, S.H. (1998). Microencapsulation of islets in PEG-amine modified alginate-poly(L-lysine)-alginate microcapsules for constructing bioartificial pancreas. *J. Ferment. Bioeng.*, Vol.86, No.2, pp.185-190, ISSN 0922-338X

Chen, H., Teramura, Y., Iwata, H. (2011). Co-immobilization of urokinase and thrombomodulin on islet surfaces by poly(ethylene glycol)-conjugated phospholipid. *J. Control. Release.* Vol.150, No.2, pp.229-234, ISSN 0168-3659

Chluba, J., Voegel, J.C., Decher, G., Erbacher, P., Schaaf, P., & Ogier, J. (2001). Peptide hormone covalently bound to polyelectrolytes and embedded into multilayer architectures concerving full biological activity. *Biomacromolecules*, Vol.2, No.3, pp. 800-805, ISSN 1525-7797

Chow, L.W., Wang, L.J., Kaufman, D.B., & Stupp, S.I. (2010). Self-assembling nanostructures to deliver angiogenic factors to pancreatic islets. *Biomaterials*, Vol.31, No.24, pp.6154-6161, ISSN 0142-9612

Contreras, J. L., Smyth, C.A., Bilbao, G., Young, C.J., Thompson, J.A., & Eckhoff, D.E. (2002). 17, beta-estradiol protects isolated human pancreatic islets against proinflammatory cytokine-induced cell death: Molecular mechanisms and islet functionality. *Transplantation*, Vol.74, No.9, pp.1252–1259, ISSN 0041-1337

Cui, W., Barr, G., Faucher, K.M., Sun, X.-L., Safley, S.A., Weber, C.J., & Chaikof, E.L. (2004). A Membrane-mimetic barrier for islet encapsulation. *Transplant. Proc.*, Vol.36, No.4, pp.1206-1208, ISSN 0041-1345

Dai, Z., Wilson, J.T., & Chaikof, E.L. (2007). Construction of pegylated multilayer architectures via (strept)avidin/biotin interactions. *Materials science and engineering C*, Vol.27, No.3, pp.402-408, ISSN 0928-4931

Daoud, J., Petropavlovskaia, M., Rosenberg, L., & Tabrizian, M. (2010). The effect of extracellular matrix components on the preservation of human islet function in vitro. *Biomaterials*, Vol.31, No.7, pp.1676-1682, ISSN 0142-9612

Dawson, R.M., Broughton, R.L., Stevenson, W.T., & Sefton, M.V. (1987). Microencapsulation of CHO cells in a hydroxyethyl methacrylate-methyl methacrylate copolymer. *Biomaterials*, Vol.8, No.5, pp.360-366, ISSN 0142-9612

Decher, G., & Schlenoff, J.B. (Eds.). (2002). *Multilayer Thin Films: Sequential Assembly of Nanocomposite Materials*, Wiley-VCH, ISBN 3-527-30440-1, Weinheim.

Dembczynski, R., Jankowski, T. (2001). Determination of pore diameter and molecular weight cut-off of hydrogel-membrane liquid-core capsules for immunoisolation. *J. Biomater. Sci. Polym. Ed.*, Vol.12, No.9, pp.1051-1058, ISSN 0920-5063

De Koker, S., De Geest, B.G., Cuvelier, C., Ferdinande, L., Deckers, W., Hennink, W.E., De Smedt, S., Mertens, N. (2007). In vivo cellular uptake, degradation, and biocompatibility of polyelectrolyte microcapsules. *Adv. Funct. Mater.*, Vol.17, No.18, pp.3754-3763, ISSN 1616-301X

De Vos, P., De Haan, B., & Van Schilfgaarde, R. (1997). Effect of the alginate composition on the biocompatibility of alginate-polylysine microcapsules. *Biomaterials*, Vol.18, No.3, pp.273-278, ISSN 0142-9612

De Vos, P., Smedema, I., van Goor, H., Moes, H., van Zanten, J., Netters, S., de Leij, L.F.M., de Haan, A., & de Haan, B.J. (2003). Association between macrophage activation and function of micro-encapsulated rat islets. *Diabetologia*, Vol.46, No.5, pp.666-673, ISSN 0012-186X

De Vos, P., Van Hoogmoed, C.G., Van Zanten, J., Netter, S., Strubbe, J.H., & Busscher, H.J. Long-term biocompatibility, chemistry, and function of microencapsulated pancreatic islets. (2003). *Biomaterials*, Vol. 24, No.2, pp.305–312, ISSN 0142-9612

De Vos, P., Van Straaten, J.F.M., Nieuwenhuizen, A.G., De Groot, M., Ploeg, R.J., De Haan, B.J., & Van Schilfgaarde, R. (1999). Why do microencapsulated islet grafts fail in the absence of fibrotic overgrowth? *Diabetes*, Vol.48, No.7, pp.1381-1388, ISSN 0012-1797

Dor, Y., Brown, J., Martinez, O.I., & Melton, D.A. (2004). Adult pancreatic beta-cells are formed by self-duplication rather than stem-cell differentiation. *Nature*, Vol.429, No.6987, pp41-46, ISSN 0028-0836

Doxastakis, M., Sum, A.K., & de Pablo, J.J. (2005). Modulating Membrane Properties: The Effect of Trehalose and Cholesterol on a Phospholipid Bilayer. *J. Phys. Chem. B*, Vol.109, No.50, pp.24173–24181, ISSN 1520-6106

Drucker, D. (2001). Development of glucagon-like peptide-1-based pharmaceuticals as therapeutic agents for the treatment of diabetes. *Current Pharmaceutical Design*, Vol.7, No.14, pp.1399-1412, ISSN 1381-6128

Drucker, D. (2002). Biological actions and therapeutic potential of the glucagon-like peptides. *Gastroenterology*, Vol.122, No.2, pp.531-544, ISSN 0016-5085

Esmon, C.T. (2004) Crosstalk between inflammation and thrombosis. *Maturitas*, Vol.47, No.4, pp.305–314, ISSN 0378-5122

Esmon, N., Owen, W., & Esmon, C.T. (1982). Isolation of a membrane-bound cofactor for thrombin-catalyzed activation of protein C. *J. Biol. Chem.*, Vol.257, No.2, pp.859–864, ISSN 0021-9258

Gimi, B., Kwon, J., Kuznetsov, A., Vachha, B., Magin, R.L., Philipson, L.H., & Lee, J.B. (2009). A nanoporous, transparent microcontainer for encapsulated islet therapy. *J. Diabetes Sci. Technol.*, Vol.3, No.2, pp.297-303, ISSN 1932-2968

Godbey, W.T., Wu, K.K., & Mikos, A.G. (1999). Size matters: molecular weight affects the efficiency of poly(ethylenimine) as a gene delivery vehicle. *J. Biomed. Mater. Res.*, Vol.45, No.3, pp.286-275, ISSN 0021-9304

Hallé, J.P., Leblond, F.A., Pariseau, J.F., Jutras, P., Brabant, M.J., & Lepage, Y. (1994). Studies on small (< 300 microns) microcapsules: II. Parameters governing the production of alginate beads by high voltage electrostatic pulses. *Cell Transplant*, Vol.3, No.5, pp.365-372, ISSN 0963-6897

Haug, A., Larsen, B., & Smidsrød, O. (1974). Uronic acid sequence in alginate from different sources. *Carbohydrate Research*, Vol.32, No.2, pp.217–225, ISSN 0008-6215

Herring, B.J., Kandaswamy, R., Harmon, J.V., Ansite, J.D., Clemmings, S.M., Sakai, T., Paraskevas, S., Eckman, P.M., Sageshima. J., Nakano, M., Sawada, T., Matsumoto, I., Zhang, H.J., Sutherland, D.E., & Bluestone, J.A. (2004). Transplantation of cultured islets from two-layer preserved pancreases in type I diabetes with anti-CD3 antibody. *Am. J. Transplant.* Vol.4, No.3, 390-401, ISSN 1600-6135

Holz, G.G., & Chepurny, O.G. (2003). Glucagon-like peptide-1 synthetic analogs: new therapeutic agents for use in the treatment of diabetes mellitus. *Curr. Med. Chem.*, Vol.10, No.22, pp.2471-2483, ISSN 0929-8673

Hortin, G.L., Lok, H.T., & Huang, S.T. (1997). Progress toward preparation of universal donor red cells. Artif. Cells Blood Subs. *Immobil. Biotechnol.*, Vol.25, No.5, pp.487-491, ISSN 1073-1199

Hsu, B.R., Chen, H.C., Fu, S.H., Huang, Y.Y., & Huang, H.S. (1994). The use of field effects to generate calcium alginate microspheres and its application in cell transplantation. *J. Formos. Med. Assoc.*, Vol.93, No.3, pp.240-245, ISSN 0929-6646

Hume, P.S., Bowman, C.N., & Anseth, K.S. (2011). Functionalized PEG hydrogels through reactive dip-coating for the formation of immunoactive barriers. *Biomaterials*, Vol.32, No.26, pp.6204-6212, ISSN 0142-9612

Ichii, H., Sakuma, Y., Pileggi, A., Fraker, C., Alvarez, A., Montelongo, J., Szust, J., Khan, A., Inverardi, L., Naziruddin, B., Levy, M.F., Klintmalm, G.B., Goss, J.A., Alejandro, R., & Ricordi, C. (2007). Shipment of human islets for transplantation. *Am. J. Transplant.*, Vol.7, No.4, pp.1010-1020, ISSN 1600-6135

Inui, O., Teramura, Y., & Iwata, H. (2010). Retention dynamics of amphiphilic polymers PEG-lipids and PVA-Alkyl on the cell surface. *ACS Appl. Mater. Interfaces*, Vol.2, No.5, pp.1514-1520, ISSN 1944-8244

Iwata, H., Takagi, T., Amemiya, H., Shimizu, H., Yamashita, K., Kobayashi, K., & Akutsu, T. (1992). Agarose for a bioartificial pancreas. J. Biomed. Mater. Res., Vol.26, No.7, pp.967-977, ISSN 0021-9304

Jeong, J.-H., Hong, S.W., Hong, S., Yook, S., Jung, Y., Park, J.-B., Khue, C.D., Im, B.-H., Seo, J., Lee, H., Ahn, C.-H., Lee, D.Y., & Byun, Y. (2011). Surface camouflage of pancreatic islets using 6-arm-PEG-catechol in combined therapy with tacrolimus and anti-CD154 monoclonal antibody for xenotransplantation. *Biomaterials*, Vol.32, No.31. pp.7961-7970, ISSN 0142-9612

Johansson, U., Elgue, G., Nilsson, B., & Korsgren, O. (2005). Composite islet-endothelial cell grafts: a novel approach to counteract innate immunity in islet transplantation. *Am. J. Transplant.*, Vol.5, No.11, pp.2632-2639, ISSN 1600-6135

Kharlampieva, E., & Sukhishvili, S.A. (2006). Hydrogen-bonded layer-by-layer films. *Polymer Reviews*, Vol.46, No.4, pp.377-395, ISSN 1558-3724

Kim, S.C., Han, D.J., Kim, I.H., Woo, K.O., We, Y.M., Kang, S.Y., Back, J.H., Kim, Y.H., Kim, J.H., & Lim, D.G. (2005). Comparative study on biologic and immunologic characteristics of the pancreas islet cell between 24 degrees C and 37 degrees C culture in the rat. *Transplant. Proc.*, Vol.37, pp.3472-3475, ISSN 0041-1345

Kim, T.G., & Park, T.G. (2006). Biomimicking extracellular matrix: Cell Adhesive RGD Peptide modified electrospun poly(D,L-lactic-co-glycolic acid) nanofiber mesh. *Tissue Eng.*, Vol.12, No.2, pp.221-233, ISSN 1076-3279

King, A., Andersson, A., & Sandler, S. (2000). Cytokine-induced functional suppression of microencapsulated rat pancreatic islets in vitro. *Transplantation*, Vol.70, No.2, pp.380-383, ISSN 0041-1337

King, A., Sandler, S., Andersson, A. (2001). The effect of host factors and capsule composition on the cellular overgrowth on implanted alginate capsules. *J. Biomed. Mater. Res.*, Vol.57, No.3, pp.374-383, ISSN 0021-9304

Kizilel, S., Scavone, A., Liu, X., Nothias, J.M., Ostrega, D., Witkowski, P., & Millis, M. (2010). Encapsulation of pancreatic islets within nano-thin functional polyethylene glycol coatings for enhanced insulin secretion. *Tissue Eng. A*, Vol.16, No.7, pp.2217-2228, ISSN 1937-3341

Koster, K.L., Lei, Y.P., Anderson, M., Martin, S., & Bryant, G. (2000). Effects of vitrified and nonvitrified sugars on phosphatidylcholine fluid-to-gel phase transitions. *Biophys. J.*, Vol.78, No.4, pp.1932-1946, ISSN 0006-3495

Kozlovskaya, V., & Sukhishvili, S.A. (2006). pH-Controlled permeability of layered hydrogen-bonded polymer capsules. *Macromolecules*, Vol.39, No.16, pp.5569-5572, ISSN 0024-9297

Krishnamurthy, V.R., Wilson, J.T., Cui, W., Song, X., Lasanajak, Y., Cummings, R.D., & Chaikof, E.L. (2010). Chemoselective Immobilization of Peptides on Abiotic and Cell Surfaces at Controlled Densities. *Langmuir*, Vol.26, No.11, pp.7675-7678, ISSN 0743-7463

Krol, S., Guerra, S., Grupillo, M., Diasporo, A., Gliozzi, A., & Marchetti, P. (2006). Multilayer nanoencapsulation. New approach for immune protection of human pancreatic islets. *Nano Lett.*, Vol.6, pp.1933-1939, ISSN 1530-6984

Lacy, P., & Kostianovsky, M. (1967) Method for the isolation of intact islets of Langerhans from the rat pancreas. *Diabetes*, Vol.16, No.1, pp.35-39, ISSN 0012-1797

Lee, D.Y., Nam, J.H., & Byun, Y. (2004). Effect of polyethylene glycol grafted onto islet capsules on prevention of splenocytes and cytokine attacks. *J. Biomater. Sci. Polym. Ed.* Vol.15, No.6, pp.753–766, ISSN 0920-5063

Lee, D. Y., Nam, J.H., Byun, Y. (2007). Functional and histological evaluation of transplanted pancreatic islets immunoprotected by PEGylation and cyclosporine for 1 year. *Biomaterials*, Vol.28, No.11, pp.1957–1966, ISSN 0142-9612

Lee, D.Y., Park, S.J., Nam, J.H., & Byun, Y. (2006). A new strategy toward improving immunoprotection in cell therapy for diabetes mellitus: long-functioning PEGylated islets in vivo. *Tissue Eng.*,Vol.12,No.3, pp.615–623, ISSN 1076-3279

Lee, D.Y., Park, S.J., Nam, J.H., & Byun, Y. (2006). A combination therapy of PEGylation and immunosuppressive agent for successful islet transplantation. *J. Control. Release*, Vol.110, No.2, pp.290–295, ISSN 0168-3659

Lee, D.Y., Yang, K., Lee, S., Chae, S.Y., Kim, K.W., Lee, M.K., Han, D.J., & Byun, Y. (2002). Optimization of monomethoxy-polyethylene glycol grafting on the pancreatic islet capsules. *J. Biomed. Mater. Res.*, Vol.62, No.3, pp.372–377, ISSN 0021-9304

Lee, S.-H., Lee, S., Youn, Y.S., Na, D.H., Chae, S.Y., Byun, Y., & Lee, K.C. (2005). Synthesis, Characterization, and Pharmacokinetic Studies of PEGylated Glucagon-like Peptide-1. *Bioconjug. Chem.*, Vol.16, No.2, pp.377-382, ISSN 1043-1802

Lim, F., & Sun, A.M. (1980). Microencapsulated islets as bioartificial endocrine pancreas. *Science*, vol.210, No.4472, pp.908-910, ISSN 0036-8075

Lin, C.-C., & Anseth, K.S. (2009). Glucagon-like peptide-1 functionalized PEG hydrogels promote survival and function of encapsulated pancreatic beta-cells. *Biomacromolecules*, Vol.10, No.9, pp.2460-2467, ISSN 1525-7797

Lin, C.-C., Metters, A.T., & Anseth, K.S. (2009). Functional PEG-peptide hydrogels to modulate local inflammation induced by the pro-inflammatory cytokine TNF-alpha. *Biomaterials*, Vol.30, No.28, pp.4907-4914, ISSN 0142-9612

Liu, X.Y., Nothias, J.M., Scavone, A., Garfinkel, M., & Millis, J.M. (2010). Biocompatibility investigation of polyethylene glycol and alginate-poly-L-lysine for islet encapsulation. *ASAIO J.*, Vol.56, No.3, pp.241-245, ISSN 1058-2916

Luan, N.M., Teramura, Y., & Iwata, H. (2011). Layer-by-layer co-immobilization of soluble complement receptor 1 and heparin on islets. *Biomaterials*, Vol.32, No.27, pp.6487-6492, ISSN 1878-5905

Lv, N., Song, M.Y., Kim, E.K., Park, J.W., Kwon, K.B., & Park, B.H. (2008). Guggulsterone, a plant sterol, inhibits NF-kappa B activation and protects pancreatic beta cells from cytokine toxicity. *Mol. Cell Endocrinol.*, Vol.289, No.1–2, pp.49–59, ISSN 0303-7207

Martens, P., Blundo, J. Nilasaroya, A., Odell, R.A., Cooper-White, J., & Poole-Warren, L.A. (2007). Effect of poly(vinyl alcohol) macromer chemistry and chain interactions on hydrogel mechanical properties. *Chem. Mater*, Vol.19, No.10, pp.2641-2648, ISSN 0897-4756

Meloche, R.M. (2007).Transplantation for the treatment of Type I diabetes. *World J. Gastroenterol.*, Vol.13, No.47, pp.6347-6355, ISSN 1007-9327

Menger, F.M., Seredyuk, V.A., Kitaeva, M.V., Yaroslavov, A.A., & Melik-Nubarov, N.S. (2003). Migration of poly-L-lysine through a lipid bilayer. *J. Am. Chem. Soc.*, Vol.125, No.10, pp.2846-2847, ISSN 0002-7863

Mørch, Y.A., Donati, I., Strand, B.L., & Skjåk-Braek, G. (2006). Effect of Ca2+, Ba2+, and Sr2+ on alginate microbeads. *Biomacromolecules*, Vol.7, No.5, pp.1471-1480, ISSN 1525-7797

Murdoch, T.B., Ghee-Wilson, D., Shapiro, A.M., & Lakey, J.R. (2004). Methods of human islet culture for transplantation. *Cell Transplant.*, Vol.13, No.6, pp.605-617, ISSN 0963-6897

Nafea E.H., Marson, A., Poole-Warren, L.A., & Martens, P.J. (2011). Immunoisolating semi-permeable membranes for cell encapsulation: Focus on hydrogels. *J. Controlled Release*, Vol.154, No.2, pp.110-122, ISSN 0168-3659

Narang, A.S., & Mahato, R.I. (2006). Biological and biomaterial approaches for improved islet transplantation. *Pharmacol. Rev.* Vol.58, No.2, pp.194–243, ISSN 0031-6997

Nielsen, J.H. (1982). Effect of growth hormone, prolactin, and placental lactogen on insulin content and release, and deoxyribonucleic acid synthesis in cultured pancreatic islets. *Endocrinology*, Vol.110, No.2, pp.600-606, ISSN 0013-7227

Opara, E.C., Mirmalek-Sani, S.-H., Khanna, O., Moya, M.L., & Brey, E.M. (2010). Design of a bioartificial pancreas. *J. Investigative Medicine*, Vol.58, No.7, pp.831-837, ISSN 1708-8267

Panza, J.L., Wagner, W.R., Rilo, H.L., Rao, R.H., Beckman, E.J., & Russell, A.J. (2000). Treatment of rat pancreatic islets with reactive PEG. *Biomaterials*, Vol.21, No.11, pp.1155–1164, ISSN 0142-9612

Paulick, M.G., Forstner, M.B., Groves, J.T., & Bertozzi, C.R. (2007). A chemical approach to unraveling the biological function of the glycosylphosphatidylinositol anchor. *Proc. Natl. Acad. Sci. U. S. A.*, Vol.104, No.51, pp.20332–20337, ISSN 0027-8424

Pi, J., Bai, Y., Zhang, Q., Wong, V., Floering, L.M., Daniel, K., Reece, J.M., Deeney, J.T., Andersen, M.E., Corkey, B.E., & Collins, S. (2007). Reactive oxygen species as a signal in glucose-stimulated insulin secretion. *Diabetes*, Vol.56, No.7, pp.1783-1791, ISSN 0012-1797

Pratt, J.R., Hibbs ,M.J., Laver, A.J., Smith, R.A., & Sacks, S.H. (1996). Effects of complement inhibition with soluble complement receptor -1 on vascular injury and inflammation duringrenal allograft rejection in the rat. *Am. J. Pathol.*, Vol.149, No.6, pp.2055-2066, ISSN 0002-9440

Qi, Z., Shen, Y., Yanai, G., Yang, K., Shirouzu, Y., Hiura, A., & Sumi, S. (2010). The in vivo performance of polyvinyl alcohol macro-encapsulated islets. *Biomaterials*, Vol.31, No.14, pp.4026-4031, ISSN 0142-9612

Rabuka, D., Forstner, M.B., Groves, J.T., & Bertozzi, C.R. (2008). Noncovalent cell surface engineering: incorporation of bioactive synthetic glycopolymers into cellular membranes. *J. Am. Chem. Soc.*, Vol.130, No.18, pp.5947–5953, ISSN 0002-7863

Raymond, M.-C., Neufeld, R.J., Poncelet, D., (2004). Encapsulation of brewers yeast in chitosan coated carrageenan microspheres by emulsification/thermal gelation. *Artif. Cells, Blood Substit. Immobil. Biotechnol.*, Vol.32, No.2, pp.275-291, ISSN 1073-1199

Riachy, R., Vandewalle, B., Conte, J.K., Moerman, E., Sacchetti, P., Lukowiak, B., Gmyr,V., Bouckenooghe, T., Dubois, M., & Pattou, F. (2002). 1,25-dihydroxyvitamin D3 protects RINm5F and human islet cells against cytokine-induced apoptosis: Implication of the antiapoptotic protein A20. *Endocrinology*, Vol.143, No.12, pp.4809–4819, ISSN 0013-7227

Ricordi, C., Lacy, P.E., Sterbenz, K., & Davie, J.M. (1987). Low-temperature culture of human islets or in vivo treatment with L3T4 antibody produces a marked prolongation of islet human-to mouse xenograft survival. *Proc. Natl. Acad. Sci. USA*, Vol.84, No.22, pp.8080-8084, ISSN 0027-8424

Ricordi, C., & Strom, T.B. (2004) Clinical islet transplantation: advances and immunological challenges. *Nat. Rev. Immunol.*, Vol. 4, No.4, pp.259-268, ISSN 1474-1733

Ris, F., Hammar, E., Bosco, D., Pilloud, C., Maedler, K., Donath, M.Y., Oberholzer, J., Zeender, E., Morel, P., Rouiller, D., & Halban, P.A. (2002). Impact of integrin-matrix matching and inhibition of apoptosis on the survival of purified human beta-cells in vitro. *Diabetologia*, Vol.45, pp.841-850, ISSN 0012-186X

Robertson, R.P. (2000) Successful islet transplantation for patients with diabetes - Fact or fantasy? *N. Engl. J. Med.*, Vol.343, No.4, pp.289-290, ISSN 0028-4793

Robertson, R.P., & Harmon, J.S. (2007). Pancreatic islet beta-cell and oxidative stress: the importance of glutathione peroxidase. *FEBS Lett.*, Vol.581, No.19, pp.3743-3748, ISSN 0014-5793

Sakai, S., Mu, C., Kawabata, K., Hashimoto, I., & Kawakami, K. (2006). Biocompatibility of subsieve-size capsules versus conventional-size microcapsules. *J. Biomed. Mater. Res. A*, Vol.78, No.2, pp.394-398, ISSN 1549-3296

Schneider, S., & Klein, H.H. (2011). Preserved insulin secretion capacity and graft function of cryostored encapsulated rat islets. *Regul. Pept.*, Vol.166, No.1-3, pp.135-138, ISSN 0167-0115

Scott, M.D., & Murad, K.L. (1998). Cellular camouflage: fooling the immune system with polymers. *Curr. Pharm. Des.*, Vol. 4, No.6, pp.423–438, ISSN 1381-6128

Stabler, C.L., Sun, X.L., Cui, W., Wilson, J.T., Haller, C.A., & Chaikof, E.L. (2007). Surface re-engineering of pancreatic islets with recombinant azido-thrombomodulin. *Bioconjug. Chem.*, Vol.18, No.6, pp.1713-1715, ISSN 1043-1802

Stabler, C., Wilks, K., Sambanis, A., & Constantinidis, I. (2001). The effects of alginate composition on encapsulated betaTC3 cells. *Biomaterials*, Vol.22, No.11, pp.1301-1310, ISSN 0142-9612

Stein, E., Mullen, Y., Benhamou, P.Y., Watt, P.C., Hober, C., Watanabe, Y., Nomura, Y., & Brunicardi, F.C. (1994). Reduction in immunogenicity of human islets by 24 degrees C culture. *Transplant. Proc.*,Vol.26, No.2, pp.755, ISSN 0041-1345

Stosic-Grujicic, S., Maksimovic, D., Badovinac, V., Samardzic, T., Trajkovic, V., Lukic, M., & Stojkovic, M.M. (2001). Antidiabetogenic effect of pentoxifylline is associated with systemic and target tissue modulation of cytokines and nitric oxide production. *J. Autoimmun.* Vol.16, No.1, pp.47–58, ISSN 0896-8411

Strand, B.K., Ryan, L., Veld, P.I., Kulseng, B., Pokstad, A.M., Skjak-Braek, G., & Espevik, T. (2001). Poly-L-lysine induces fibrosis on alginate microcapsules via the induction of cytokines. *Cell Transplant.*, Vol.10, No.3, pp.263-277, ISSN 0963-6897

Su, J., Hu, B.H., Lowe, W.L.Jr., Kaufman, D.B., & Messersmith. P.B. (2010). Anti-inflammatory peptide-functionalized hydrogels for insulin-secreting cell encapsulation. *Biomaterials*, Vol.31, No.2, pp.308-314, ISSN 0142-9612

Takemoto, N., Teramura, Y., & Iwata, H. (2011). Islet surface modification with urokinase through DNA hybridization. *Bioconjug. Chem.*, Vol.22, No.4, pp.673-678, ISSN 1043-1802

Tang, Z., Wang, Y., Podsiadlo, P., & Kotov, N.A. (2006). Biomedical Applications of Layer-by-Layer Assembly: From Biomimetics to Tissue Engineering. *Adv. Mater.*, Vol.18, No.24, pp.3203-3224, ISSN 0935-9648

Teramura, Y., & Chen, H. (2010). Control of cell attachment through polyDNA hybridization. *Biomaterials*, Vol.31, No.8, pp.2229-2235, ISSN 0142-9612

Teramura, Y., & Iwata, H. (2008). Islets surface modification prevents blood-mediated inflammatory responses. *Bioconjug. Chem.*, Vol.19, No.7, pp.1389-1395, ISSN 1043-1802

Teramura, Y., & Iwata, H. (2009). Islet encapsulation with living cells for improvement of biocompatibility. *Biomaterials*, Vol.30, No.12. pp.2270-2275, ISSN 0142-9612

Teramura, Y., Kaneda, Y., & Iwata, H. (2007). Islet-encapsulation in ultra-thin layer-by-layer membranes of poly(vinyl alcohol) anchored to poly(ethylene glycol)-lipids in the cell membrane. *Biomaterials*, Vol.28, No.32, pp.4818-4825, ISSN 0142-9612

Teramura, Y., Kaneda, Y., Totani, T., & Iwata, H. (2008). Behavior of synthetic polymers immobilized on a cell membrane. *Biomaterials*, Vol.29, No.10, pp.1345-1355, ISSN 0142-9612

Teramura, Y., & Minh, L. (2010). Microencapsulation of islets with living cells using polyDNA-PEG-lipid conjugate. *Bioconjug. Chem.*, Vol.21, No.4 pp.792-796, ISSN 1043-1802

Thu, B., Bruheim, P., Espevik, T., Smidsrød, O., Soon-Shiong, P., & Skjåk-Braek, G. (1996). Alginate polycation microcapsules. II. Some functional properties. *Biomaterials*, Vol.17, No.11, pp.1069-1079, ISSN 0142-9612

Tierney, L. M., McPhee, S.J., & Papadakis, M.A. (2002). *Current medical diagnosis & treatment*. (International edition), Large Medical Books/McGraw-Hill, ISBN: 007-1376-88-7, New York.

Wang, R.N., & Rosenberg, L. (1999). Maintenance of beta-cell function and survival following islet isolation requires re-establishment of the islet-matrix relationship. *J. Endocrinol.*, Vol.163, No.2, pp.181-190, ISSN 0022-0795

Weber, L.M., & Anseth, K.S. (2008). Hydrogel encapsulation environments functionalized with extracellular matrix interactions increase islet insulin secretion. *Matrix Biology*, Vol.27, No.8, pp.667-673, ISSN 0945-053X

Weber, L.M., Cheung, C.Y., & Anseth, K.S. (2007). Multifunctional pancreatic islet encapsulationbarriers achieved via multilayer PEG hydrogels. *Cell Transplant.*, Vol.16, No.10, pp.1049-1057, ISSN 0963-6897

Weber, L.M., Hayda, K.N., Haskins, K., & Anseth, K.S. (2007). The effects of cell-matrix interactions on encapsulated beta-cell function within hydrogels functionalized with matrix-derived adhesive peptides. *Biomaterials*, Vol.28, No.19, pp.3004-3011, ISSN 0142-9612

Weber, L.M., He, J., Bradley, B., Haskins, K., & Anseth, K.S. (2006). PEG-based hydrogels as an in vitro encapsulation platform for testing controlled beta-cell microenvironments. *Acta Biomaterialia*, Vol.2, No.1, pp.1-8, ISSN 1742-7061

Weber, L.M., Lopez, C.G., & Anseth, K.S. (2008). Effects of PEG hydrogel crosslinking density on protein diffusion and encapsulated islet survival and function. *J. Biomed. Mater. Res. A*, Vol.90, No.3, pp.720-729, ISSN 1549-3296

Wilson, J.T., & Chaikof, E.L. (2008). Challenges and emerging technologies in the immunoisolation of cells and tissues. *Adv. Drug Deliv. Rev.*, Vol.60, No.2, pp.124-145, ISSN 0169-409X

Wilson, J.T., Cui, W., & Chaikof, E.L. (2008). Layer-by-layer assembly of a conformal nanothin PEG coating for intraportal islet transplantation. *Nano Lett.*, Vol.8, No.7, pp.1940-1948, ISSN 1530-6984

Wilson, J.T., Cui, W., Kozlovskaya, V., Kharlampieva, E., Pan, D., Qu, Z., Krishnamurthy, V.R., Mets, J., Kumar, V., Wen, J., Song, Y., Tsukruk, V.V., & Chaikof, E.L. (2011). Cell surface engineering with polyelectrolyte multilayer thin films. *J. Am. Chem. Soc.*, Vol.133, No.18, pp.7054-7064, ISSN 0002-7863

Wilson, J.T., Krishnamurthy, V.R., Cui, W., Qu, Z., & Chaikof, E.L. (2009). Noncovalent cell surface engineering with cationic graft copolymers. *J. Am. Chem. Soc.*, Vol.131, No.51, pp.18228-18229, ISSN 0002-7863

Wolters, G.H.J., Fritschy, W.M., Gerrits, D., & Vanschilfagaarde, R. (1992). A versatile alginate droplet generator applicable for microencapsulation of pancreatic islets. *J. Appl. Biomat.*, Vol.3, No.4, 281-286. ISSN: 1045-4861

Wyman, J.L., Kizilel, S., Skarbek, R., Zhao, X., Connors, M., Dillmore, W.S., Murphy, W.L., Mrksich, M., Nagel, S.R., & Garfinkel, M.R. (2007). Immunoisolating pancreatic islets by encapsulation with selective withdrawal. *Small*, Vol.3, No.4, pp.683-690, ISSN 1613-6810

Xie, D., Smyth, C.A., Eckstein, C., Bilbao, G., Mays, J., Eckhoff, D.E., & Contreras, J.L. (2005) Cytoprotection of PEG-modified adult porcine pancreatic islets for improved xenotransplantation. *Biomaterials*, Vol.26, No.4, pp.403-412, ISSN 0142-9612

Yamamoto, T., Mita, A., Ricordi, C., Messinger, S., Miki, A., Sakuma, Y., Timoneri, F., Barker, S., Fornoni, A., Molano, R.D., Inverardi, L., Pileggi, A., & Ichii, H. (2010). Prolactin supplementation to Culture Medium improves beta-cell survival. *Transplantation*, Vol.89, No.11, pp.1328-1335, ISSN 0041-1337

Yang, H., Iwata, H., Shimizu, H., Takagi, T., Tsuji, T., & Ito, F. (1994). Comparative studies of in vitro and in vivo function of three different shaped bioartificial pancreases made of agarose hydrogel. *Biomaterials*, Vol.15, No.2, pp.113-120, ISSN 0142-9612

Zhang, X., He, H., Yen, C., Ho, W., Lee, L.J. (2008). A biodegradable, immunoprotective, dual nanoporous capsule for cell-based therapies. *Biomaterials*, Vol.29, No.31, 4253-4259, ISSN 0142-9612

Zhi, Z.L., Liu, B., Jones, P.M., & Pickup, J.C. (2010). Polysaccharide multilayer nanoencapsulation of insulin-producing beta-cells grown as pseudoislets for potential cellular delivery of insulin. *Biomacromolecules*, Vol.11, No.3, pp.610-616, ISSN 1525-7797

Zimmermann, U., Mimietz, S., Zimmermann, H., Hillgartner, M., Schneider, H., Ludwig, J., Hasse, C., Haase, A., Rothmund, M., & Fuhr, G. (2000). Hydrogel-based non-autologous cell and tissue therapy. *Biotechniques*, Vol.29, No.3, pp.564-572, ISSN 0736-6205

Zimmermann, U., Thurmer, F., Jork, A., Weber, M., Mimietz, S., Hillgärtner, M., Brunnenmeier, F., Zimmermann, H., Westphal, I., Fuhr, G., Nöth, U., Haase, A., Steinert, A., & Hendrich, C. (2001). A novel class of amitogenic alginate microcapsulesfor long-term immunoisolated transplantation. *Ann. N. Y. Acad. Sci.*, Vol.944, pp.199-215, ISSN 0077-8923

2

In-Situ Forming Biomimetic Hydrogels for Tissue Regeneration

Rong Jin
Institute of Nanochemistry and Nanobiology, Shanghai University, Shanghai, P.R. China

1. Introduction

Tissue loss or organ failure caused by injury or damage is one of the most serious and costly problems in human health care. Tissue engineering, proposed by Langer *et al.* in the early 1990's (Langer & Vacanti, 1993), is an emerging strategy of regenerative biomedicine that holds promise for the restoration of defect tissues and organs. The concept of tissue engineering is defined as "the application of the principles and methods of engineering and the life sciences towards the fundamental understanding of structure-function relationships in normal and pathological mammalian tissues and the development of biological substitutes that restore, maintain or improve tissue function" (Langer & Vacanti, 1993). In order to accomplish these goals by tissue engineering, three essential components are required, that is, cells for the generation of new tissues, scaffolds for supporting the cell growth and the regeneration of new tissues, and bioactive factors capable of stimulating biological signals *in vivo* for cell proliferation, dfferentiation and tissue growth. Among these, the scaffolds play an important role in the success of tissue regeneration since they serve as temporary temples to mimick the excellular matrix for cell growth and interim mechanical stability for tissue regneration and integration,.

Hydrogels are one of most used bio-scaffolds in the field of tissue enginereering. They are three-dimensional, water-swollen, crosslinked networks of hydrophilic polymers. Wichterle and Lim for the first time reported on hydrogels based on the hydroxyethyl methacrylate (HEMA) for biological use in 1960 (Wichterle & Lim, 1960). Due to their unique tissue-like properties, such as high water content and good permeability to oxygen and metabolites, hydrogels have been widely studied as biomimetic extracellular matrixes for tissue regeneration. Hydrogels may be used by implantation or injection, which corresponds to so-called preformed hydrogels or in-situ forming hydrogels. From the clinical point of view, in-situ forming hydrogels are highly desirable since they gain advantages over preformed hydrogels: (1) Enabling minimally invasive surgeries for implantation; (2) Formation in any desired shape in good alignment with surrounding tissue defects; (3) Easy encapsulation of bioactive molecules and progenic cells. Therefore, in-situ forming hydrogels have received much attention in recent years.

2. Strategies to design in-situ forming hydrogels

In-situ forming hydrogels are referred to as hydrophilic polymer networks that are in-situ formed in the body after the injection of liquid gel precursors. They are typically categorized into chemical hydrogels and physical hydrogels according to the mechanism underlying the network formation. Chemical hydrogels are those that are prepared by chemically covalent crosslinking of polymers. On the other hand, physical hydrogels are obtained by physical interactions of polymers, such as stereocomplex formation, hydrophobic interactions, and ionic interactions. So far, different crosslinking methods have been developed to prepare in-situ forming hydrogels, which are described in detail as follows.

2.1 Chemical crosslinking

Chemical crosslinking produces irreversible, also called permanent hydrogels. Generally, the hydrogels have robust mechanical properties and chemcial stability, which are favorable as supportive scaffolds for tissue engineering. Furthermore, covalent cross-linking is a good means to precisely control the cross-linking density of chemical hydrogels, thus controlling the hydrogels properties such as degradation time and mechanical strength.

2.1.1 Michael-type addition

Michael-type addition reaction is one of the commonly used approaches for the preparation of hydrogels, especially, in-situ forming hydrogels. By this approach, in-situ forming hydrogels can be obtained by mixing aqueous solutions of polymers bearing nucleophilic (amine or thiol) and electrophilic groups (vinyl, acrylate or maleimide) (Mather et al., 2006). For example, Feijen *et al.* prepared in-situ forming hydrogels based on vinyl sulfone-conjugated dextran and poly(ethylene glycol) (PEG) thiols through Michael-type addition (Hiemstra et al., 2007a). The gelation times can be tailored from 7 to 0.5 min when the drgree of vinyl sulfone subsititution increased from 4 to 13. Additionally, by varying the degrees of substitution, dextran molecular weights and polymer concentrations, the storage moduli of the hydrogels can be adjusted from 3 to 46 kPa, and the degradation time from 3 to 21 days. In another study, Hubbell et al. reported on smart hydrogels that were formed in-situ by the addition of thiol-containing oligopeptides to multi-arm vinyl sulfone-terminated PEG (Fig. 1) (Lutolf et al., 2003). Rheology test showed the pH condition plays an important role in the gel formation. With the increasing pH value from 7 to 8, the gelation time decreased from 24 to 4 min. Also, different thiol-bearing peptides (e.g., cysteine-bearing peptides) could be

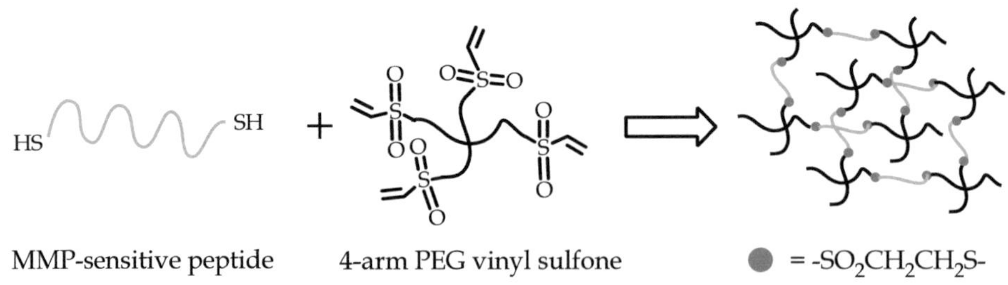

Fig. 1. In-situ forming cell-responsive hydrogels prepared from vinyl sulfone-functionalized 4-arm poly(ethylene glycol) and the MMP-sensitive bis-cysteine peptide.

incorporated to yield biofunctional or bioreponsive hydrogels, enhancing cell adhesion and matrix production (Seliktar et al., 2004). This indicates that Michael-type addition is an ideal method for the preparation of in-situ crosslinked hybrid hydrogels.

2.1.2 Radical polymerization

Radical polymerization is one of the most frequently used crosslinking methods to prepare robust and stable in-situ forming hydrogels. Radicals are created from initiator molecules through thermal, redox or photointiated mechanisms. Then, the radicals propagate through unreacted double bonds during polymerization to form long kinetic chains, and the chains react further with each other to form crosslinked polymeric networks (Ifkovits & Burdick, 2007). In general, macromers bearing vinyl groups are relatively biocompatible and more favourable as compared to monomers. Since the reaction takes place in aqueous solutions, the conversion of double bonds is high due to the high mobility of reacting species during gel formation. This also decreases the potential toxicity of the materials. PEG and PEG-based copolymers are commonly used synthetic biomaterials (Nguyen & West, 2002) (Fig. 2). They can be functionalized with acryl chloride and further crosslinked in the presence of free-radical initiators under a physiological environment to form hydrogels. Multi-arm polymers such as 4-arm and 8-arm PEG were also employed to increase the crosslinking density because of their increased functionality as compare to linear analogues. Other types of polymers are natural polymers such as dextran, hyaluronic acid and collagen (Dong et al., 2005; S.H. Kim et al., 1999; Y.D. Park et al., 2003). As compared to synthetic polymers, they have different functional groups (hydroxyl, amine or carboxylic groups) on their polymer backbones and are amenable to various chemical modifications. The number of double bonds introduced can be precisely controlled on demand. Thus, the properties of free-radical polymerized hydrogel such as gelation time, mechanical properties and degradation profiles can be adjusted for use in different tissue engineering.

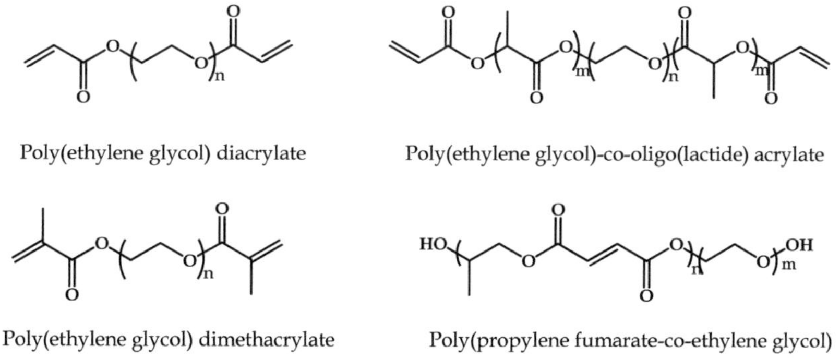

Fig. 2. Commonly used poly(ethylene glycol)-based polymers for in-situ forming hydrogels

2.1.3 Enzymatic crosslinking

In-situ hydrogels formation using enzymes have emerged recently. Enzymes are known to exhibit a high degree of substrate specificity, which potentially avoids side reactions during crosslinking. Another advantage of the enzymatic crosslinking is of mild gelation conditions (e.g. physiological conditions), favourable for tissue regeneration.

Horseradish peroxidase (HRP) has been recently employed in the preparation of in situ forming hydrogels. HRP is a single-chain b-type hemoprotein that catalyzes the coupling of phenols or aniline derivatives in the presence of hydrogen peroxide (Kobayashi et al., 2001). Crosslinking reaction takes place via a carbon-carbon bond at the ortho positions and/or via a carbon-oxygen bond between the carbon atom at the ortho position and the phenoxy oxygen in the phenol moieties (Fig. 3a). For example, Feijen *et al.* reported on HRP-mediated in-situ forming dextran-tyramine hydrogels for cartilage tissue engineering (Jin et al., 2007). Tyramine was conjugated to dextran by first activation of the hydroxyl groups in dextran using p-nitrophenyl chloroformate and then treatment with tyramine by aminolysis. The gelation rates induced by enzyme-mediated crosslinking can be readily adjusted from minutes to seconds by varying the HRP concentrations. By the same approach, Jin and Lee *et al.* prepared the chitosan-phloretic acid and hyaluronic acid-tyramine hydrogels for cartilage tissue engineering (Jin et al., 2009) and protein delivery applications (F. Lee et al., 2009), respectively. The disadvantage of this approach is the use of hydrogen peroxide. It is reported that high concentration of hydrogen peroxide (>0.2 mM) may induce cell apoptosis (Asada et al., 1999). Therefore, it is important to control the amount of hydrogen peroxide used in a cell-favourable range.

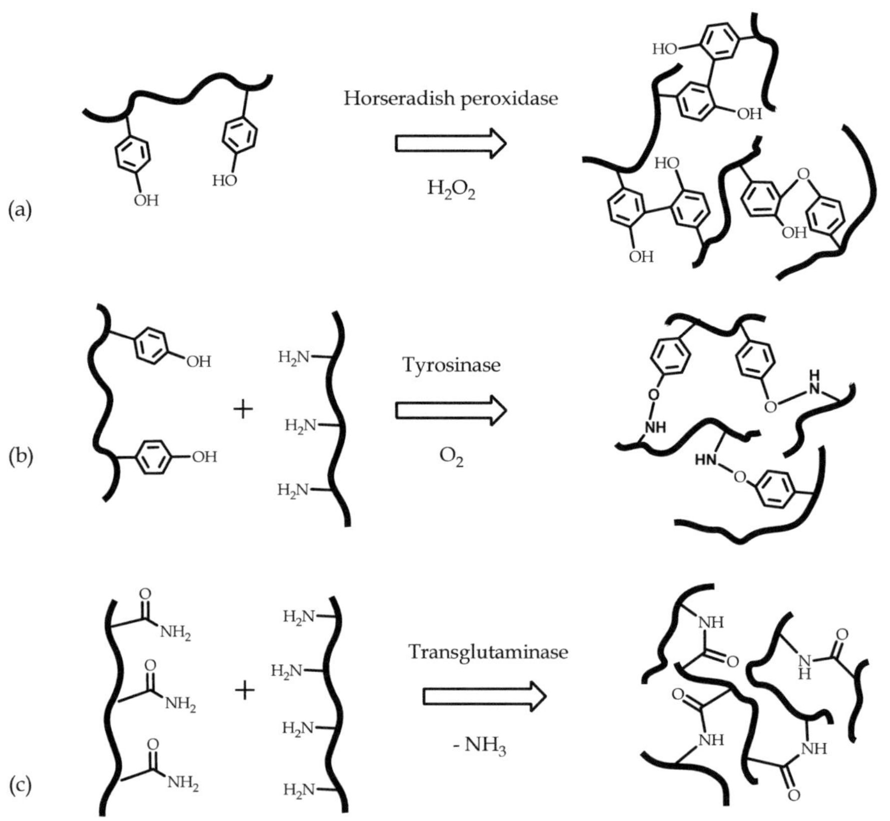

Fig. 3. Enzymatic crosslinking method to prepare in-situ forming hydrogels

Tyrosinase is another enzyme used to form in-situ forming hydrogels. Unlike HRP, the tyrosinase crosslinks phenol-containing polymers in the presence of oxygen instead of hydrogen peroxide (Fig. 3b). Besides, tyrosinase is an oxidative enzyme present in the animal or human body. These features imply milder gelation conditions and better cyto-biocompatibility of tyrosinase-crosslinked hydrogels as compared to HRP-crosslinked hydrogels. So far, only few studies have been conducted to construct hydrogels using tyrosinase. For example, Payne *et al.* reported on the hydrogels from the composites of chitosan and gelatin (Chen et al., 2003). The strength of tyrosinase-catalyzed gels could be adjusted by altering the gelatin and chitosan compositions. The author speculated that the gel system may be useful as emergency dressings for burns and wounds.

Transglutaminase (TGase) is an enzyme frequently used in protein crosslinking. Recently, it has been employed in the preparation of in-situ forming hydrogels. TGase catalyzes an acyl-transfer reaction between the γ-carboxamide group of protein bound glutaminyl residues and the amino group of ε-lysine residues, resulting in the formation of ε-(γ-glutamyl)lysine isopeptide side chain bridges (Sperinde & Griffith, 1997) (Fig. 3c). McHale *et al.* designed and synthesized engineered elastin-like polypeptide (ELPs) hydrogels that are capable of undergoing enzyme-initiated gelation via tissue TGase (McHale et al., 2005). Two kinds of polymer solutions ELP[KV6-112] and ELP[QV6-112] were first mixed and the gels were formed within an hour after enzymes and $CaCl_2$ were subsequently added. In another study, Sanborn *et al.* investigated TGase-catalyzed gelation of peptide-modified PEG and the results showed that gelation times ranged from 9 to 30 min (Sanborn et al., 2002). To shorten the gelation time, Messersmith *et al.* attempted to rationally design the peptide substrates by increasing their specificity (Hu & Messersmith, 2003). It was found that the introduction of an N-terminal L-3,4-dihydroxylphenylalanine (DOPA) residue into the tripeptides resulted in ca. 2.4-fold increment in specificity. This facilitated the gel formation with shorter gelation time of 2 min.

2.1.4 Peptide ligation

Peptide ligation is often employed in the synthesis of proteins and enzymes, which is based on chemoselective reaction of two unprotected peptide segments. Recently, this reaction was explored by Grinstaff *et al* for the preparation of in-situ forming hydrogels due to the mild chemical reaction conditions. A typical peptide ligation reaction is based on the reaction of aldehyde groups in poly(ethylene glycol) derivatives and NH_2-terminal cysteine moieties in peptide dendrons, which can form thiazolidine rings (Wathier et al., 2004) (Fig.4a). The gelation process took place within a few minutes. However, these hydrogels were intact for short periods of time (about 1 week) due to the reversible thiazolidine ring formation. To overcome the problem, the same group developed stable hydrogels prepared from poly(ethylene glycol) with endcapped ester-aldehyde groups instead of aldehyde groups (Wathier et al., 2006) (Fig.4b). The ester-aldehyde groups firstly reacted with the NH_2-terminal cysteine moieties to form the thiazolidine ring, which then can undergo a rearrangement to give chemically stable pseudoproline ring. The mechanical properties of the hydrogels depend on the concentrations of the polymer solutions and different ratios of aldehyde to cysteine reactive functionality. Degradation studies demonstrated that the pseudoproline ring was more stable than the thiazolidine ring and the hydrogels retained their shape and size with less than 10% weight loss for more than 6 months.

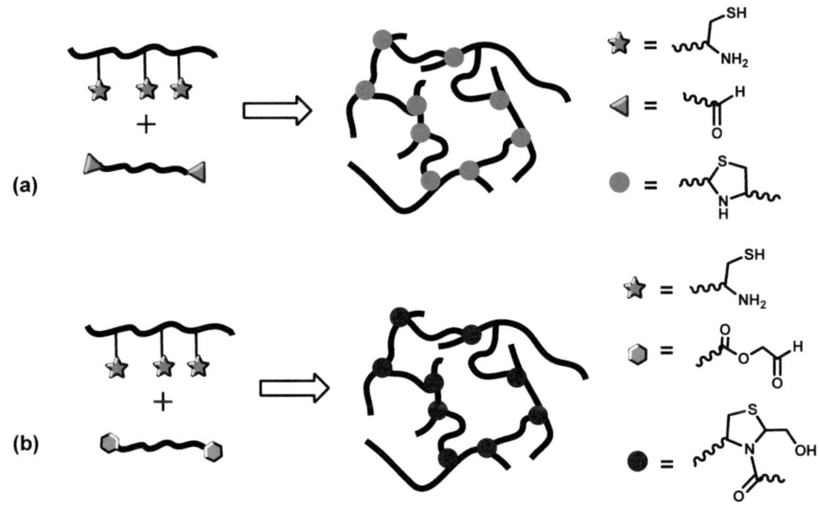

Fig. 4. Preparation of in-situ forming hydrogel via peptide ligation

2.2 Physical crosslinking

Much attention has been paid in the preparation of in-situ forming physical hydrogels. The advantages of physical crosslinking are relatively good biocompatibility and less toxicity since toxic crosslinking reagent or initiators are not used during crosslinking. However, physically-crosslinked hydrogels are gnerally unstable and mechanically weak. The changes in the environment such as pH, temperature and ionic strength may lead to the disruption of the gel network. Typical physical crosslinking methods include stereocomplexation, hydrophobic interactions and ionic interaction.

2.2.1 Stereocomplexation

A typical polymer used for stereocomplex formation is poly(lactide) (PLA). It is a kind of aliphatic polyesters, which are known to be biocompatible and render PLA-based hydrogels biodegradable. Lactide has three possible configurations, which refer to D-lactide, L-lactide and meso-lactide according to the arrangement of substituents around the chiral carbon. The corresponding polymers are defines as poly(L-lactide) (PLLA), poly(D-lactide) (PDLA) and poly(D,L-lactide) (PDLLA). The formation of stereocomplexes when mixing PLLA and PDLA was first reported by Ikada *et al.* (Ikada et al., 1987). The stereocomplexation not only occurs in the blends of PLLA and PDLA homopolymers, but also in water-soluble PLA and poly(ethylene glycol) (PEG) block copolymers, such as linear (Hiemstra et al., 2005) and multiarm PEG-PLLA and PEG-PDLA block copolymers (Hiemstra et al., 2006). This gives the possibility to design different PLA-conjugated materials in hydrogel preparation. For example, Hennink *et al.* reported on physical hydrogels based on the stereocomplexation of PLA-dextran conjugates (de Jong et al., 2000). L- and D-lactic acid oligomers were coupled to dextran to yield dex-(L)lactate and dex-(D)lactate. It was found that the degree of polymerization of lactic acid oligomers must be at least 11 to obtain the hydrogels. The stereocomplex crosslinking can be detected by X-ray diffraction (de Jong et al., 2002). Varying the degree of polymerization of oligomer, the degree of substitution of dex-lactate and the water content of dex-lactate solutions, the properties of the hydrogels can be well modulated.

2.2.2 Hydrophobic interaction

The hydrophobic interaction provides another driving force of physical gelation. Some amphiphilic copolymers can undergo a sol-gel transition via this mechanism. Typical amphiphilic polymers are block copolymers based on poly(ethylene oxide) (PEO) and poly(propylene oxide) (PPO), also called as Pluronics® (Fusco et al., 2006). It was found that, when increasing temperature, the PEO-PPO polymers can undergo a dehydration process. This leads to the formation of hydrophobic domains and, in turn, transition of an aqueous liquid to a hydrogel. The drawbacks of PEO-PPO hydrogels include rapid erosion, potential cytotoxicity (Khattak et al., 2005), and non-biodegradability. Alternatively, polyester-based copolymers that are biodegradable received much attention. For example, Jeong and co-workers described thermosensitive, biodegradable hydrogels based on poly(ethylene oxide) and poly(lactic acid) (Jeong et al., 1997). Solutions of the diblock copolymers were shown to be in solution state at 45°C, but gel state at body temperature. However, the encapsulation of drugs at an elevated temperature might lead to denaturation of bioactive agents such as therapeutic proteins or growth factors. The same group subsequently reported on a series of triblock copolymers of poly(ethylene oxide) and poly(lactic acid)/poly(glycolic acid) (Jeong et al., 1999). This thermosensitive hydrogel system is inverse to the hydrogel based on poly(ethylene oxide)-co-poly(lactic acid) diblock polymers, that is, poly(ethylene oxide)-b-(D,L-lactic acid-co-glycolic acid)-b-poly(ethylene oxide) triblock copolymers were found to be in a solution at room temperature, but form a hydrogel when the temperature is increased to 37°C. This makes the gel system easy to handle and favourable for tissue engineering applications (Jeong et al., 2000). Moreover, the sol–gel transition temperature and degradation properties can be adjusted by the polymer concentration, molecular weight of poly(ethylene oxide) and the lactic acid/glycolic acid ratio in the poly(lactic acid-co-glycolic acid) blocks. In another study, Ding *et al.* reported that the end groups have a surprising effect on the hydrogel formation (L. Yu et al., 2006). The results showed that the transition temperature increased with a decreasing hydrophobicity of the end groups. Importantly, it is noted that sol-gel transition takes place only when the hydrophobic interactions are strong enough to induce the large-scale self-assembly of micelles. However, over hydrophobicity and higher temperature lead to precipitation of polymers as a result of the break of micelle structure.

Polypeptide or peptide-conjugated polymers is another type of polymers that can be used to form hydrogels via hydrophobic interactions. The formation of hydrogels from polypeptides are based on coil-coil interaction, triggered by the self-assembly of peptide sequences. The coiled-coil interaction is one of the basic folding patterns of native proteins and consists of two or more helices winding together to form a superhelix (Y.B. Yu, 2002). A series of hydrogels based on peptides or peptide/synthtic polymer hydrids were made. Typical examples are synthetic N-(2-hydroxypropyl)methacrylamide (HPMAm) copolymer grated with coiled-coil protein motifs (C. Wang et al., 1999; J. Yang et al., 2006b). The gelation time can be adjusted by the length and the number of coiled-coil grafts per chain an ranged from a few minutes to several days (J. Yang et al., 2006a). Besides, it was found that at least 4 heptads were needed to achieve hydrogels formation. In another study, Xu *et al.*

reported on the hydrogels based on genetically engineered protein block copolymers with 2 coiled-coil domains in a random coil polyelectrolyte (Xu & Kopeček, 2008). The self-assmebly process between coil-coils was influenced by the protein concetration, pH and temperature. Changes in the peptide sequence of the coil-coil domains endow hydrogels with different stability.

2.2.3 Ionic interaction

Ionic interaction is another route to construct in-situ forming hydrogels. For example, the hydrogels can be formed by ionic interactions between water-soluble charged polymers and their di- or multi-valent counter-ions. As a typical example, alginate, a naturally occurring polysacchride, can form a hydrogel network in the presence of calcium ions under physiological conditions. The mechanism underlying the ionic crosslinking is ion exchange of sodium ions by calcium ions in the carboxylic groups and subsequent formation of an egg-box structure (Gombotz & Wee, 1998). The hydrogel is degradable slowly with the diffusion of calcium ions out of the hydrogels and finally excreted from the kidney. Fur ther studies showed that alginate-based hydrogels with $CaSO_4$ usually reveal a heterogeneous structure due to the difficulty in the control of gelation kinetics (Kuo & Ma, 2001). This phenomenon also occured for the hydrogels using $CaCl_2$ (Skjak-Brvk et al., 1989). In contrast, $CaCO_3$ can give homogeneous alginate hydrogel, while its low solubility is unfavourable for further biomedical application. Ma et al. reported on crosslinked alginate hydrogels using $CaCO_3$-GDL (D-glucono-d-lactone) and $CaSO_4$-$CaCO_3$-GDL systems (Kuo & Ma, 2001). Gelation rates and mechanical properties of the alginate hydrogels could be controlled by varying the composition of calcium compound systems and alginate concentration, thereby giving rise to structurally uniform hydrogels.

In-situ forming hydrogels can also be prepared by ionic interactions between polycations and polyanions. For example, Hennink et al. reported on self-gelling hydrogels based on oppositely charged dextran microspheres (Tomme et al., 2005). These charged dextran-microspheres were prepared by radical polymerization of hydroxyethyl methacrylate-derivatized dextran (dex-HEMA) with methacrylic acid (MAA) or dimethylaminoethyl methacrylate (DMAEMA). Hydrogels could be formed as a result of ionic interactions between oppositively-charged microspheres. The networks of hydrogels were disrupted either by applied stress, low pH or high ionic strength. Reversible yield point from rheological analysis indicated that this hydrogel system can be applied for controlled delivery of pharmaceutically active proteins and tissue engineering. However, a main disadvantage of ionically-crosslinked hydrogels is that their mechanical strength is far from satisfactory when they are served as scaffolds for tissue regeneration.

3. Biomimetic hydrogels

The success of tissue engineering depends on biomimetic hydrgoel scaffolds that possess controlled structures and on-demand properties to modulate specific cellular behaviors. The development of suitable synthetic methods encompassing chemistry and molecular biology open a new way for the design of biomimetic hydrogels mimicking basic processes of living systems. In general, biomimetic hydrogels can be categorized into bioactive, bioresponsive, and biofunctional hydrogels.

3.1 Bioactive hydrogels

Much effort has been made in the design of bioactive hydrogels which can instruct cell behaviors and promote tissue regeneration. A well-know bioactive ligand is cell-adhesive peptides, e.g., Arg-Gly-Asp (RGD). It was revealed that RGD-modified PEG diacrylate hydrogels could induce enhanced cell attachment and mineralized matrix deposition of osteoblasts as compared to RGD-free hydrogels (Burdick & Anseth, 2002). Natural proteins such as collagen and its analogs may also serve as bioactive ligands due to inherent nature of biological recognition. Seliktar *et al.* reported on the preparation of proteins (collagen, albumin and fibrinogen) conjugated with acrylated PEG and subsequent hydrogel formation by photopolymerization (Gonen-Wadmany et al., 2007) (Fig. 5). The modified protein maintained its cell-adhesive properties and supported proteolytic degradability based on the specific characteristics of the protein backbone. In another study, Lee and coworkers reported on the collagen mimetic peptide-conjugated poly(ethylene glycol) hydrogels (H.J. Lee et al., 2006). The collagen mimetic peptide (CMP) with a specific amino acid sequence, -(Pro-Hyp-Gly)x-, forms a triple helix conformation that resembles the native protein structure of natural collagens. CMP was first conjugated with acrylated PEG, which copolymerized with poly(ethylene oxide) diacrylate to create a novel PEG hydrogel. The modified protein can maintain their cell-adhesive properties and support proteolytic degradability based on the specific characteristics of the protein backbone. The biochemical analysis showed that chondrocytes-encapsulated hydrogels revealed an 87% increase in glycosaminoglycan content and a 103% increase in collagen content compared to that of control PEG hydrogels after 2 weeks. These results indicate that the CMP enhances the tissue production of cells encapsulated in the PEG hydrogel by providing cell-manipulated crosslinks and collagen binding sites that simulate natural extracellular matrix.

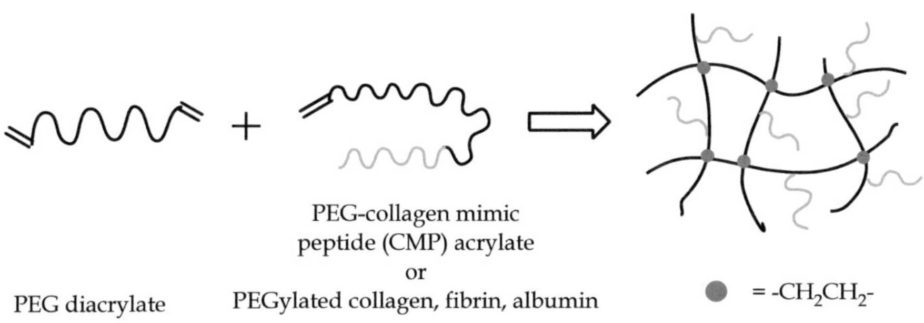

Fig. 5. Bioactive hydrogels prepared from poly(ethylene glycol) and proteins/peptide.

3.2 Bioresponsive hydrogel

Biomimetic hydrogels can response to biological components, such as enzymes, receptors and antibodies. After the hydrogels undergo a macroscopic transition (gelation, enzymatic degradation and swelling/shrinkage), this in turn directly leads to microscopic response of living cells (cell migration, differentiation, cell division and matrix production). For example, Lutolf *et al.* developed cell-responsive hydrogels that can degrade in response to local protease activity such as matrix metalloproteinase (MMP) at the cell surface. MMP is a protease family extensively involved in tissue development and remodeling. The hydrogel systems were made from vinyl sulfone-functionalized multiarmed PEG and the bis-cysteine peptide crosslinker which contained the sequence sensitive to matrix metalloproteinases (Lutolf et al., 2003). The hydrogels were proteolytically degraded via the invasion of primary human fibroblasts. The invasion process depended on MMP substrate activity, adhesion ligand concentration, and network crosslinking density. By mimicking the MMP-mediated invasion of the natural provisional matrix, the hydrogels were shown to assist tissue regeneration. These results indicate potential applications of the cell-responsive hydrogels in tissue engineering and regenerative medicine.

3.3 Biofunctional hydrogels

Mechanical modulus is in the range of 10 kPa–350 MPa for soft tissues and 10 MPa–30 GPa for hard tissues (S. Yang et al., 2001). Depending on intended application, hydrogel should provide sufficient mechanical strength so as to protect seeded cells and developing neo-tissue as well as to withstand the physiologic load. However, most of hydrogels reported so far are not qualified especially for bone or cartilage tissue regeneration due to the lack of a high mechanical strength. Thus, robust hydrogels have been developed with on-demand mechanical properties. Mechanical moduli of hydrogels are generally increasing with increasing crosslinking density. Two types of robust hydrogel systems can be classified.

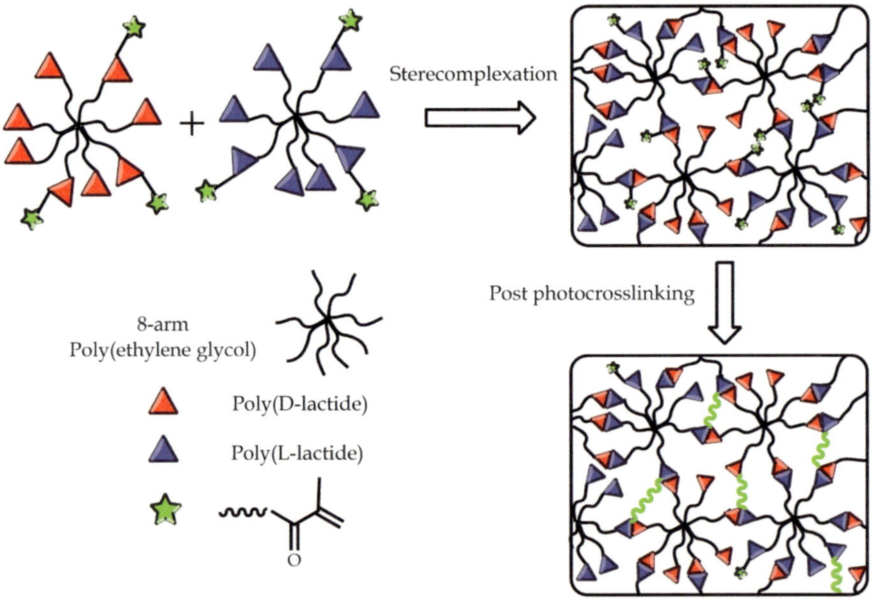

Fig. 6. Hydrogel prepared via the combination of stereocomplexation and photocrosslinking

First, hydrogels can be fabricated by double crosslinking methods. By this approach, the hydrogels have increased crosslinking density, thus improving the mechanical properties without comprising other properties such as permeability and biocompatibility. For example, Feijen and coworkers reported on the in situ hydrogels crosslinked by combining stereocomplexation and photopolymerization (Hiemstra et al., 2007b). Stereocomplexed hydrogels were first formed upon mixing solutions of an 8-arm PEG-PLLA and an 8-arm PEG-PDLLA whichh are partly functionalized with methacrylate groups (40%). These hydrogels can be postcrosslinked by UV-irradiation (Fig. 6). These double-crosslinked hydrogels showed increased mechanical moduli and prolonged degradation times compared to the hydrogels that were formed only by stereocomplexation. The photopolymerization takes place at much lower initiator concentrations (0.003 wt%) than conventional photocrosslinking systems (0.05 wt%), which greatly reduces the possibility of heating effects that can damage cells.

Second, robust hydrogels are produced that consist of two interpenetrated polymeric networks. The hydrogels with double networks contain a subset of interpenetrating networks (IPNs) formed by two hydrophilic networks, one highly crosslinked, the other loosely crosslinked. The double network structure can be obtained by pre- and post-crosslinking through exploiting the disparity of their reaction times. For example, a double netwok composed of two mechanically weak hydrophilic networks based on N, N-dimethylacrylamide and glycidyl methacrylated hyaluronan, provides a hydrogel with outstanding mechanical properties (Weng et al., 2008). Hydrogels containing more that 90% water possessed a compressive modulus and a fracture stress over 0.5 MPa and 5.2 MPa, respectively, demonstrating both hardness and toughness. Besides, it is found that both the concentrations of monomers and crosslinkers are important parameters related to the mechanical strength of double network gels. Therefore, it is easy to control the mechincal properties such as hardness and toughness independently by adjusting the compositions of of the gels for practical applications.

4. Tissue engineering applications

4.1 Cartilage tissue regeneration

Cartilage is a flexible, connective tissue in which chondrocytes are sparsely distributed in the extracellular matrixes rich in proteoglycans (PGs) and collagen fibers. Cartilage has a limited capacity for self-repair due to its avascular nature and low mitotic activity of chondrocytes. In articular cartilage, chondrocytes are the only cell type and responsible for the synthesis and maintenance of resilient extracellular matrix. Chondrocytes may undergo a dedifferentiation process during monolayer culturing and lose their phenotype. However, once cultured in hydrogels, dedifferentiated chondrocytes are able to redifferentiate (Benya & Shaffer, 1982), as indicated by their rounded morphology and the production of ECM molecules such as type II collagen and sulfated glycosaminoglycans.

4.1.1 Factors influencing cartilage regeneration

In-situ forming hydrogels enable a perfect match with irregular cartilage defects and good alignment with the surrounding tissues. Therefore, they are promising materials that can function as scaffolds for chondrocyte culturing and cartilage regeneration. Several factors may influence the cell viability, recovery or the maintenance of the chondrocytic phenotype, and correspondingly play an important role in cartilage tissue engineering.

Chemical compositions of hydrogels have been studied to explore their influence on cartilage regeneration. For example, Elisseeff *et al.* studied the cellular toxicity of transdermal photopolymerization on chondrocytes (Elisseeff et al., 1999). There was a significant decrease in the cell viability when the initiator concentration was increased from

0.012% to 0.036% or higher. In another study, Chung *et al.* noticed that a higher macromer concentration potentially compromised cell viability and growth (Chung et al., 2006). Besides, a higher polymer concentration also resulted in a decreased accumulation of matrix components such as proteoglycans and collagen type II (Sontjens et al., 2006).

Recent studies showed that the degradation properties of the gels may have a significant influence on the matrix production and distribution as well. Degradable hydrogels induced a more homogenous distribution of GAG than non-degradable hydrogels (Bryant & Anseth, 2002, 2003; Bryant et al., 2003; Martens et al., 2003). However, in fast degrading hydrogels void spaces are generally present before new matrix formation has taken place (Bryant & Anseth, 2003; Martens et al., 2003). Therefore, the degradation rate of hydrogels needs to be tailored by the combination of degradable main chain linkages and crosslinks.

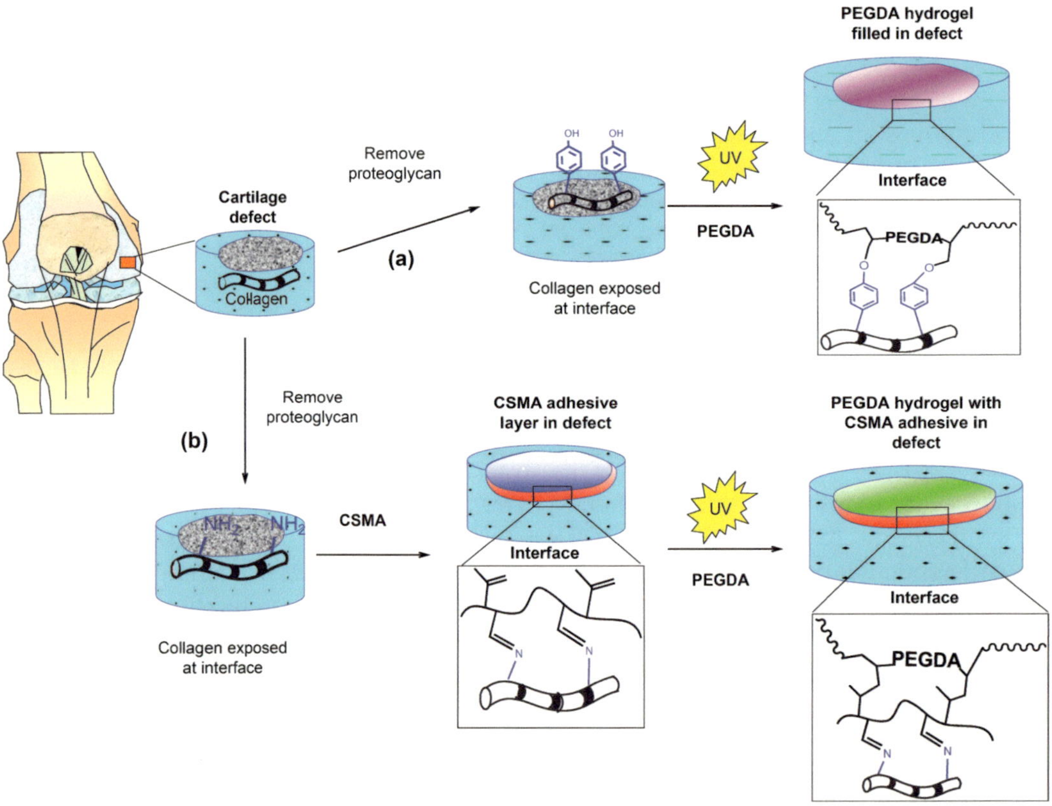

Fig. 7. Hydrogel-cartilage integration by (a) tissue-initiated photopolymerization or (b) Schiff-base formation.

A major problem for cartilage regeneration is poor integration of neocartilage with native cartilage tissue. To solve this problem, the gel precursor molecules were modified with functional groups that can react with collagen type II, a molecule present in native cartilage. For example, improved tissue adhesion and integration was achieved by tissue-initiated polymerization between acrylate groups in polymerizable PEGDA macromers and tyrosine groups in collagen when exposed to light and an oxidative reagent like H_2O_2 (D.A. Wang et al., 2007) (Figure 7a). In another approach, methacrylated chondroitin sulfate (CSMA) was functionalized to endow aldehyde moieties which was covalently attached to collagen via Schiff-base formation (D.A. Wang et al., 2004) (Figure 7b). The CSMA layer was further polymerized by photo-crosslinking of PEGDA to give a gel/cartilage integrated scaffold.

4.1.2 Hybrid hydrogels for cartilage regeneration

Understanding of the tissue structure and composition can lead to a rational material design, targeted towards mimicking the underlying biological cues and specific chemistry of cartilage. Generally, in-situ forming hydrogels have been prepared from synthetic polymers, natural polymers or their hybrids. The latter has gained increasing attention in recent years because they combine the advantage of both synthetic and natural polymers, that is, tightly defined physical, chemical and biological properties. Table 1 lists typical examples of in-situ forming hybrid hydrogels for cartilage tissue regeneration.

Synthetic polymer	Natural moiety	Comments	Ref.
PEG	MMP-sensitive peptide	Proteolytic degradability, enhanced gene expression of type II collagen and aggrecan	(Y. D. Park et al., 2004)
PVA	Chondroitin sulfate	Balance between modulus and swelling of hydrogels, enhanced matrix production	(Bryant et al., 2005)
PEG	Collagen-mimic peptide	Retention of ECM production inside hydrogel via collagen binding, enhanced chondrogenesis	(H.J. Lee et al., 2008)
Pluronic F127	Hyaluronic acid, RGD	Improved cellular adhesion and proliferation, increased matrix production	(H. Lee & Park, 2009)
PEG	Heparin	Promoted chondrocyte proliferation, while maintaining chondrogenic nature	(M. Kim et al., 2010)

Table 1. Typical examples of in-situ forming hybrid hydrogels for cartilage regeneration

4.1.3 Growth factor

During the cartilage regeneration process growth factors play a crucial role in regulating cellular proliferation, differentiation, migration, and gene expression. Besides, they have large influences on the communication between cells and their microenvironment. A number of growth factors have been studied and include bone morphogenetic protein (BMP), transforming growth factor (TGF), insulin-like growth factor (IGF) and basic fibroblast growth factor (bFGF). Their main functions in cartilage regeneration are summarized in Table 2. For example, the BMP family can stimulate mitosis and matrix production by chondrocytes and induce chondrogenesis of mesenchymal cells, triggering them to differentiate and maintain a chondrogenic phenotype (Yuji et al., 2004). TGF-β not only enhances chondrocyte proliferation, but also increases the synthesis of proteoglycans (H. Park et al., 2005).

Growth factor	Function	Ref.
BMP	Inducing chondrogenesis; Stimulating cartilage formation	(Y. Park et al., 2005)
TGF-β	Regulation of cell proliferation and differentiation; Stimulating production of proteoglycans and other matrix components	(H. Park et al., 2005)
IGF	Promotion of cartilage tissue formation	(Elisseeff et al., 2001)
bFGF	Potent modulator of cell proliferation, motility, differentiation, and survival; initiation of chondrogenesis	(K.H. Park & Na, 2008)

Table 2. Delivery of growth factors using injectable hydrogels for cartilage regeneration

4.1.4 *In vivo* studies

Many in-situ forming hydrogels so far prepared for cartilage tissue engineering have been studied in vitro, however, only a few reports have appeared on their performance in vivo. Before clinical application, in-situ forming hydrogels need to be systematically evaluated in animal models. Early *in vivo* studies generally focused on the ability of cell-seeded injectable hydrogels to generate cartilaginous tissue after subcutaneous implantation/injection in mouse models. For example, Elisseeff *et al.* implanted chondrocyte-incorporated PEO-based hydrogels, and showed that the chondrocytes survived during the photopolymerization process and proliferated without any sign of necrosis (Elisseeff et al., 1999). However, partial chondrocyte dedifferentiation and undesired fibro-cartilaginous tissue formation were observed in these gels after 6 weeks' in vivo. To retain the chondrocyte phenotype and improve cartilage regeneration, ECM components and bioactive molecules were incorporated into hydrogels. Na *et al.* showed that cartilage-specific ECM production was significantly higher in poly(N-isopropylacrylamide-co-hydroxyethyl methacrylate) hydrogels containing HA and TGF-β3 compared to those without the growth factor or HA (K.H. Park & Na, 2008). Recent studies have been directed to in-situ forming hydrogels for cartilage regeneration in animal models like rabbit and goat. Hoemann *et al.* tested the residence of in-situ forming hydrogels in rabbit joints and showed that in-situ forming chitosan/glycerol phosphate gels could reside at least 1 day in a full-thickness chondral defect, and at least 1 week in a mobile osteochondral defect (Hoemann et al., 2005). Liu et al. described osteochondral defect repair in a rabbit model using a synthetic ECM composed of hyaluronic acid and gelatin (Liu et al., 2006). At 12 weeks, the defects were completely filled with elastic, firm, translucent cartilage and showed good integration of the repair tissue with the surrounding cartilage. A summary of in vivo studies on in-situ forming hydrogels for cartilage regeneration is presented in Table 3.

Hydrogel (+/-cell)	Animal model	Outcome	Ref.
Chitosan-GP-glucosamine (+ primary calf chondrocytes)	Subcutaneous injection in nude mice	Chitosan gels supported cartilage matrix accumulation by cells 48 days after injection.	(Hoemann et al., 2005)
Hyaluronic acid (+ swine auricular chondrocytes)	Subcutaneous implantation in nude mice	Neocartilage was produced and evenly distributed in the gels after 6 weeks. The P0 and P1 chondrocytes produced neocartilage tissue that resembled native auricular cartilage after 12 weeks.	(Chung et al., 2006; Ifkovits & Burdick, 2007)
PEGDA with methacrylated chondroitin sulfate as adhesive	Goat chondral defects	Defects treated with chondroitin sulfate adhesive and hydrogel showed improved cartilage repair compared to an empty, untreated defect after 6 months.	(D.A. Wang et al., 2007)
Chitosan-GP-glucosamine	Rabbit (osteo)chondral defects	Chitosan gel can reside at least 1 day in a full-thickness chondral defect and for at least 1 week in a mobile osteochondral defect.	(Hoemann et al., 2005)
Hyaluronic acid-gelatin-PEGDA (+ autologous MSC)	Rabbit osteochondral defect	Defects were completely filled with elastic, firm, translucent cartilage at 12 weeks and showed superior integration of the repair tissue with the cartilage.	(Liu et al., 2006)
Elastin-like polypeptide	Goat chondral defects	ELP formed stable, well-integrated gels and supported cell infiltration and matrix synthesis 3 months after injection. These hydrogels degraded rapidly.	(Nettles et al., 2008)

Table 3. In-situ forming hydrogels for *in vivo* cartilage regeneration

4.2 Would healing

Wound healing is a complicated process which requires coordination of complex cell and biomaterial interactions. Desirable properties of biomaterials involve formability in situ from aqueous solutions, good adhesion to tissues at one surface (tissue surface) and resistance to adhesion to the other (free surface), and degradability without induction of inflammation (Hubbell, 1996). In-situ forming hydrogels are attractive biomaterials in the application for would healing due to their ability of adjusting the moisture of the wound tissue (wetting the dehydrate tissue and absorb exudation) and conformability of the dressing on wounds (Jones & Vaughan, 2005).

4.2.1 Chitosan-based hydrogels

Polysaccharides, e.g. chitosan, represent a class of hydrogels used as would healing materials. Native chitosan has low solubility above pH 6. Modifications on chitosan can improve its solubility and make it suitable as in-situ forming materials. Ono et al. reported potocrosslinked chitosan as a dressing for wound occlusion (Ono et al., 2000). The modified chitosan (Az-CH-LA) containing both lactose moieties and azide groups exhibited a good solubility at neutral pH. Application of ultraviolet light (UV) irradiation to Az-CH-LA produced an insoluble hydrogel within 60 s. The results showed that the chitosan hydrogels could completely stop bleeding from a cut mouse tail within 30 s and firmly adhere two pieces of sliced skins of mouse to each other. The wound healing efficacy of hydrogel was evaluated in experimental full-thickness-round wounds of skin using a mouse model and it is showed that chitosan hydrogels could significantly induce wound contraction and accelerate wound closure and healing in both db and db+ mice (Ishihara et al., 2002). Incorporation of fibroblast growth factor-2 further accelerated wound closure in db mice, however, not in db+ mice (Obara et al., 2003).

4.2.2 Alginate-gelatin hydrogels

Gelatin is a degraded form of collagen and alginate is derived from brown seaweed, both of which are biocompatible. Balakrishnan *et al.* reported on the evaluation of alginate-gelatin hydrogels for wound dressing (Balakrishnan et al., 2005). Hydrogels were prepared from oxidized alginate and gelation in the presence of borax. It was found that the hydrogel have the ability to prevent accumulation of exudates on the wound bed due to fluid uptake. In vivo experiment showed the wound covered with hydrogel was completely filled with new epithelium after two weeks using a rat model. In addition, incorporation of dibutyryl cyclic adenosine monophosphate (DBcAMP) into the in-situ forming hydrogels and sustained release of DBcAMP led to the enhancement in the rate of healing as well as re-epithelialization of the wounds (Balakrishnan et al., 2006). Complete healing was achieved within 10 days associated with mild contracture of some of the wounds.

4.3 Delivery system

From the clinical point of view, the success of a scaffold for tissue engineering is judged by its ability to regenerate tissue in both the onset and completion of tissue defect repair. During this process, the presence of growth factors in the hydrogel-based scaffolds usually helps to govern neo-tissue formation and organization. However, growth factors generally have short half-life time and are easy to lose their bioactivity(Edelman et al., 1991). Moreover, some unexpected adverse effects may occur which could be caused by initial burst release of growth factors. Therefore, the appropriate mode to deliver growth factors, make them available at the site of action and effectively controlled release of them to exert their maximum efficacy is of great important.

Direct administration of growth factors is commonly associated with problems such as a short biological half-life and easy diffusion. In-situ forming hydrogels offer significant opportunities for controlled local delivery of such biomolecules. These bioactive agents can be easily incorporated into hydrogels prior to gelation and their release kinetics can be adjusted on demand by the crosslinking density and stability of the networks. The disadvantages of in situ incorporation (Fig. 8a), however, is the potential damage of proteins during the gelation process (Sperinde & Griffith, 1997) or the occurrence of an initial burst release. To circumvent these disadvantages, growth factors can be incorporated into microparticles which can be added to the hydrogel precursor solutions (Fig. 8b). Microparticles can be prepared either from synthetic polymers (e.g. PGA, PLA, and PLGA) or from natural polymers (e.g. gelatin) (S.H. Lee & Shin, 2007). For example, Holland et al.

reported on TGF-β1-loaded-gelatin particles which were incorporated in oligo(poly(ethylene glycol) fumarate) hydrogels (Holland et al., 2003). In vitro release experiments showed a suppressed burst release and prolonged delivery of TGF-β1. Besides, when chondrocytes were embedded, an increased cellular proliferation and enhanced chondrocyte-specific gene expression was observed for the hydrogels containing TGF-loaded-gelatin particles (H. Park et al., 2005).

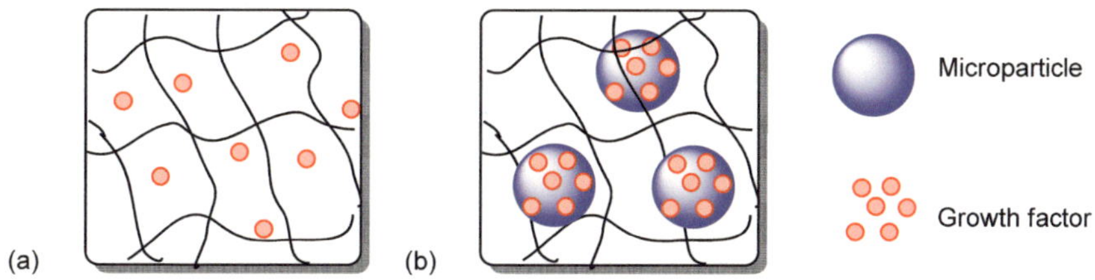

Fig. 8. Schematic representation of methods for encapsulating growth factors either by (a) direct incorporation or (b) preloading into microparticles

Another example to control the release of growth factors is to use heparin-containing hydrogels. It is known that heparin is able to bind basic fibroblast growth factor (bFGF) by the formation of a stable complex (Yayon et al., 1991). bFGF can be prevented from denaturation and proteolysis meanwhile maintaining its biological activity after its release. Besides, the use of hparin can efficiently control the release rate of bFGF. However, large quantity use of heparin induces side effects such as thrombocytopenia, thrombosis, and hemorrhage (Silver et al., 1983). To solve this problem, Prestwich et al. proposed an in-situ forming glycosaminoglycan (GAG) hydrogel based on hyaluronan (HA) and chondroitin sulfate (CS) (Cai et al., 2005). Crosslinking occur between thiol-modified HA or CS (HA-DTPH or CS-DTPH), thiol-modified heparin (HP) and poly(ethylene glycol) diacrylate (PEG-DA) to generate copolymers, containing only a small percentage of co-crosslinked thiol-modified HP, which is capable of controlled release of basic fibroblast growth factor. Notably, the diffusion of bFGF in the hydrogels can be substantially slowed down only with 1% (w/w) covalently bound heparin (relative to total glycosaminoglycan content). In vivo studies, carried out on the Balb/c mice using the hydrogels (HA-DTPH+HP-DTPH and CS-DTPH+HP-DTPH) with/without bFGF, showed that the implanted hydrogels containing bFGF enhanced the production of new blood vessels to a high extent than equal amount of injected free bFGF, indicating that covalently crosslinked HP was necessary to enhance bFGF activity and promote neovascularization.

5. Conclusion

Novel crosslinking methods provide significant opportunities for the design of in-situ forming hydrogels with multifunctional properties on demand for tissue regeneration. The fast progress in molecular biology inspires researchers to design biomimetic in situ forming hydrogels. Polymer composition and structures, hydrogel forming methods, degradation properties, mechanical strength and biocompatibility are of significant importance. Artificial extracellular matrices combining in situ forming hydrogel scaffolds, cells and growth factors hold great promise for tissue engineering, and pave the way for regenerated tissue.

6. Acknowledgment

This work was financially sponsored by the National Natural Science Foundation of China (No. 21004039), the Research Foundation for The Excellent Youth Scholars of Higher Education of Shanghai, the Scientific Research Foundation for the Returned Overseas Chinese Scholars, State Education Ministry, and the Innovative Foundation of Shanghai University.

7. References

Asada, S., Fukuda, K., Oh, M., Hamanishi, C. & Tanaka, S. (1999). Effect of Hydrogen Peroxide on the Metabolism of Articular Chondrocytes. *Inflammation Research,* Vol.48, No.7, (July 1999), pp. 399-403, ISSN 1023-3830

Balakrishnan, B., Mohanty, M., Umashankar, P.R. & Jayakrishnan, A. (2005). Evaluation of an in Situ Forming Hydrogel Wound Dressing Based on Oxidized Alginate and Gelatin. *Biomaterials,* Vol.26, No.32, (November 2005), pp. 6335-6342, ISSN 0142-9612

Balakrishnan, B., Mohanty, M., Fernandez, A.C., Mohanan, P.V. & Jayakrishnan, A. (2006). Evaluation of the Effect of Incorporation of Dibutyryl Cyclic Adenosine Monophosphate in an in Situ-Forming Hydrogel Wound Dressing Based on Oxidized Alginate and Gelatin. *Biomaterials,* Vol.27, No.8, (March 2006), pp. 1355-1361, ISSN 0142-9612

Benya, P.D. & Shaffer, J.D. (1982). Dedifferentiated Chondrocytes Reexpress the Differentiated Collagen Phenotype When Cultured in Agarose Gels. *Cell,* Vol.30, No.1, (August 1982), pp. 215-224, ISSN 0092-8674

Bryant, S.J. & Anseth, K.S. (2002). Hydrogel Properties Influence ECM Production by Chondrocytes Photoencapsulated in Poly(Ethylene Glycol) Hydrogels. *Journal of Biomedical Materials Research,* Vol.59, No.1, (January 2002), pp. 63-72, ISSN 1552-4965

Bryant, S.J. & Anseth, K.S. (2003). Controlling the Spatial Distribution of ECM Components in Degradable PEG Hydrogels for Tissue Engineering Cartilage. *Journal of Biomedical Materials Research Part A,* Vol.64A, No.1, (January 2003), pp. 70-79, ISSN 1552-4965

Bryant, S.J., Durand, K.L. & Anseth, K.S. (2003). Manipulations in Hydrogel Chemistry Control Photoencapsulated Chondrocyte Behavior and Their Extracellular Matrix Production. *Journal of Biomedical Materials Research: Part A,* Vol.67A, No.4, (June 2003), pp. 1430-1436, ISSN 1552-4965

Bryant, S.J., Arthur, J.A. & Anseth, K.S. (2005). Incorporation of Tissue-Specific Molecules Alters Chondrocyte Metabolism and Gene Expression in Photocrosslinked Hydrogels. *Acta Biomaterialia,* Vol.1, No.2, (March 2005), pp. 243-252, ISSN 1742-7061

Burdick, J.A. & Anseth, K.S. (2002). Photoencapsulation of Osteoblasts in Injectable Rgd-Modified PEG Hydrogels for Bone Tissue Engineering. *Biomaterials,* Vol.23, No.22, (November 2002), pp. 4315-4323, ISSN 0142-9612

Cai, S., Liu, Y., Shu, X.Z. & Prestwich, G.D. (2005). Injectable Glycosaminoglycan Hydrogels for Controlled Release of Human Basic Fibroblast Growth Factor *Biomaterials,* Vol.26, No.30, (October 2005), pp. 6054-6067, ISSN 0142-9612

Chen, T., Embree, H.D., Brown, E.M., Taylor, M.M. & Payne, G.F. (2003). Enzyme-Catalyzed Gel Formation of Gelatin and Chitosan: Potential for in Situ Applications. *Biomaterials,* Vol.24, No.17, (August 2003), pp. 2831-2841, ISSN 0142-9612

Chung, C., Mesa, J., Randolph, M.A., Yaremchuk, M. & Burdick, J.A. (2006). Influence of Gel Properties on Neocartilage Formation by Auricular Chondrocytes Photoencapsulated in Hyaluronic Acid Networks. *Journal of Biomedical Materials Research Part A,* Vol.77A, No.3, (June 2006), pp. 518-525, ISSN 1552-4965

de Jong, S.J., De Smedt, S.C., Wahls, M.W.C., Demeester, J., Kettenes-van den Bosch, J.J. & Hennink, W.E. (2000). Novel Self-Assembled Hydrogels by Stereocomplex Formation in Aqueous Solution of Enantiomeric Lactic Acid Oligomers Grafted to Dextran. *Macromolecules,* Vol.33, No.10, (April 2000), pp. 3680-3686, ISSN 0024-9297

de Jong, S.J., van Nostrum, C.F., Kroon-Batenburg, L.M.J., Kettenes-van den Bosch, J.J. & Hennink, W.E. (2002). Oligolactate-Grafted Dextran Hydrogels: Detection of Stereocomplex Crosslinks by X-Ray Diffraction. *Journal of Applied Polymer Science,* Vol.86, No.2, (October 2002), pp. 289-293, ISSN 0021-8995

Dong, C.-M., Wu, X., Caves, J., Rele, S.S., Thomas, B.S. & Chaikof, E.L. (2005). Photomediated Crosslinking of C6-Cinnamate Derivatized Type I Collagen. *Biomaterials,* Vol.26, No.18, (June 2005), pp. 4041-4049, ISNN 0142-9612

Edelman, E.R., Mathiowitz, E., Langer, R. & Klagsbrun, M. (1991). Controlled and Modulated Release of Basic Fibroblast Growth Factor. *Biomaterials,* Vol.12, No.7, 1991), pp. 619-626,

Elisseeff, J., Anseth, K., Sims, D., McIntosh, W., Randolph, M. & Langer, R. (1999). Transdermal Photopolymerization for Minimally Invasive Implantation. *Proceedings of the National Academy of Sciences of the United States of America,* Vol.96, No.6, (September 1999), pp. 3104-3107, ISSN 1091-6490

Elisseeff, J., McIntosh, W., Fu, K., Blunk, T. & Langer, R. (2001). Controlled-Release of IGF-I and TGF-Beta1 in a Photopolymerizing Hydrogel for Cartilage Tissue Engineering. *Journal of Orthopaedic Research,* Vol.19, No.6, (November 2001), pp. 1098-1104, ISSN 0736-0266

Fusco, S., Borzacchiello, A. & Netti, P.A. (2006). Perspectives On: PEO-PPO-PEO Triblock Copolymers and Their Biomedical Applications. *Journal of Bioactive and Compatible Polymers,* Vol.21, No.2, (March 2006), pp. 149-164, ISSN 0883-9115

Gombotz, W.R. & Wee, S.F. (1998). Protein Release from Alginate Matrices. *Advanced Drug Delivery Reviews,* Vol.31, No.3, (May 1998), pp. 267–285, ISSN 0169-409X

Gonen-Wadmany, M., Oss-Ronen, L. & Seliktar, D. (2007). Protein-Polymer Conjugates for Forming Photopolymerizable Biomimetic Hydrogels for Tissue Engineering. *Biomaterials,* Vol.28, No.26, (September 2007), pp. 3876-3886, ISSN 0142-9612

Hiemstra, C., Zhong, Z., Dijkstra, P.J. & Feijen, J. (2005). Stereocomplex Mediated Gelation of PEG-$(PLA)_2$ and PEG-$(PLA)_8$ Block Copolymers. *Macromolecular Symposium,* Vol.224, No.1, (April 2005), pp. 119-132, ISSN 1521-3900

Hiemstra, C., Zhong, Z., Li, L., Dijkstra, P.J. & Feijen, J. (2006). In-Situ Formation of Biodegradable Hydrogels by Stereocomplexation of PEG-(PLLA)8 and PEG-(PDLA)8 Star Block Copolymers. *Biomacromolecules,* Vol.7, No.10, (October 2006), pp. 2790-2795, ISSN 1525-7797

Hiemstra, C., vanderAa, L.J., Zhong, Z., Dijkstra, P.J. & Feijen, J. (2007a). Novel in Situ Forming, Degradable Dextran Hydrogels by Michael Addition Chemistry:

Synthesis, Rheology, and Degradation. *Macromolecules,* Vol.40, No.4, (January 2007a), pp. 1165-1173, ISSN 0024-9297

Hiemstra, C., Zhou, W., Zhong, Z., Wouters, M. & Feijen, J. (2007b). Rapidly in Situ Forming Biodegradable Robust Hydrogels by Combining Stereocomplexation and Photopolymerization. *Journal of the American Chemical Society,* Vol.129, No.32, (August 2007b), pp. 9918-9926, ISSN 0002-7863

Hoemann, C.D., Sun, J., Légaré, A., McKee, M.D. & Buschmann, M.D. (2005). Tissue Engineering of Cartilage Using an Injectable and Adhesive Chitosan-Based Cell-Delivery Vehicle. *Osteoarthritis and Cartilage,* Vol.13, No.4, (April 2005), pp. 318-329, ISSN 1063-4584

Holland, T.A., Tabata, Y. & Mikos, A.G. (2003). In Vitro Release of Transforming Growth Factor-[Beta]1 from Gelatin Microparticles Encapsulated in Biodegradable, Injectable Oligo(Poly(Ethylene Glycol) Fumarate) Hydrogels. *Journal of Controlled Release,* Vol.91, No.3, (September 2003), pp. 299-313, ISSN 0168-3659

Hu, B.-H. & Messersmith, P.B. (2003). Rational Design of Transglutaminase Substrate Peptides for Rapid Enzymatic Formation of Hydrogels. *Journal of the American Chemical Soiety,* Vol.125, No.47, (October 2003), pp. 14298-14299, ISSN 0002-7863

Hubbell, J.A. (1996). Hydrogel Systems for Barriers and Local Drug Delivery in the Control of Wound Healing. *Journal of Controlled Release,* Vol.39, No.2-3, (May 1996), pp. 305-313, ISSN 0168-3659

Ifkovits, J.L. & Burdick, J.A. (2007). Review: Photopolymerizable and Degradable Biomaterials for Tissue Engineering Applications. *Tissue engineering,* Vol.13, No.10, (October 2007), pp. 2369-2385, 1076-3279

Ikada, Y., K.Jamshidi, Tsuji, H. & Hyon, S.-H. (1987). Stereocomplex Formation between Enantiomeric Poly(Lactide). *Macromolecules,* Vol.20, No.4, (April 1987), pp. 904-906, ISSN 0024-9297

Ishihara, M., Nakanishi, K., Ono, K., Sato, M., Kikuchi, M., Saito, Y., Yura, H., Matsui, T., Hattori, H., Uenoyama, M. & Kurita, A. (2002). Photocrosslinkable Chitosan as a Dressing for Wound Occlusion and Accelerator in Healing Process. *Biomaterials,* Vol.23, No.3, (February 2002), pp. 833-840, ISSN 0142-9612

Jeong, B., Bae, Y.H., Lee, D.S. & Kim, S.W. (1997). Biodegradable Block Copolymers as Injectable Drug-Delivery Systems. *Nature,* Vol.388, No.1, (August 1997), pp. 860-862, ISSN 0028-0836

Jeong, B., Bae, Y.H. & Kim, S.W. (1999). Thermoreversible Gelation of PEG-PLGA-PEG Triblock Copolymer Aqueous Solutions. *Macromolecules,* Vol.32, No.21, (October 1999), pp. 7064-7069, ISSN 0024-9297

Jeong, B., Bae, Y.H. & Kim, S.W. (2000). In Situ Gelation of PEG-PLGA-PEG Triblock Copolymer Aqueous Solutions and Degradation Thereof. *Journal of Biomedical Materials Research,* Vol.50, No.2, (May 2000), pp. 171-177, ISSN 1549-3296

Jin, R., Hiemstra, C., Zhong, Z. & Feijen, J. (2007). Enzyme-Mediated Fast in Situ Formation of Hydrogels from Dextran-Tyramine Conjugates. *Biomaterials,* Vol.28, No.18, (June 2007), pp. 2791-2800, ISSN 0142-9612

Jin, R., Moreira Teixeira, L.S., Dijkstra, P.J., Karperien, M., van Blitterswijk, C.A., Zhong, Z.Y. & Feijen, J. (2009). Injectable Chitosan-Based Hydrogels for Cartilage Tissue Engineering. *Biomaterials,* Vol.30, No.13, (May 2009), pp. 2544-2551, ISSN 0142-9612

Jones, A. & Vaughan, D. (2005). Hydrogel Dressing in the Management of a Variety of Wound Types: A Review. *Journal of Orthopaedic Nursing*, Vol.9, No.1, (December 2005), pp. S1-S11, ISSN 1361-3111

Khattak, S.F., Bhatia, S.R. & Roberts, S.C. (2005). Pluronic F127 as a Cell Encapsulation Material: Utilization of Membrane-Stabilizing Agents. *Tissue engineering*, Vol.11, No.5-6, (May-June 2005), pp. 974-983, ISSN 1076-3279

Kim, M., Shin, Y., Hong, B.-H., Kim, Y.-J., Chun, J.-S., Tae, G. & Kim, Y.H. (2010). In Vitro Chondrocyte Culture in a Heparin-Based Hydrogel for Cartilage Regeneration. *Tissue Engineering Part C*, Vol.16, No.1, (February 2010), pp. 1-10, ISSN 1076-3279

Kim, S.H., Won, C.Y. & Chu, C.C. (1999). Synthesis and Characterization of Dextran-Based Hydrogel Prepared by Photocrosslinking. *Carbohydrate Polymers*, Vol.40, No.3, (November 1999), pp. 183-190, ISSN 0144-8617

Kobayashi, S., Uyama, H. & Kimura, S. (2001). Enzymatic Polymerization. *Chemical Review*, Vol.101, No.12, (November 2001), pp. 3793-3818, ISSN 0009-2665

Kuo, C.K. & Ma, P.X. (2001). Ionically Crosslinked Alginate Hydrogels as Scaffolds for Tissue Engineering: Part I. Structure, Gelation Rate and Mechanical Properties. *Biomaterials*, Vol.22, No.6, (March 2001), pp. 511-521, ISSN 0142-9612

Langer, R. & Vacanti, J.P. (1993). Tissue Engineering. *Science*, Vol.260, No.5110, (May 1993), pp. 920-926, ISSN 0036-8075

Lee, F., Chung, J.E. & Kurisawa, M. (2009). An Injectable Hyaluronic Acid-Tyramine Hydrogel System for Protein Delivery. *Journal of Controlled Release*, Vol.134, No.3, (March 2009), pp. 186-193, ISSN 0168-3659

Lee, H. & Park, T.G. (2009). Photo-Crosslinkable, Biomimetic, and Thermo-Sensitive Pluronic Grafted Hyaluronic Acid Copolymers for Injectable Delivery of Chondrocytes. *Journal of Biomedical Materials Research Part A*, Vol.88A, No.3, (March 2009), pp. 797-806, ISSN 1552-4965

Lee, H.J., Lee, J.-S., Chansakul, T., Yu, C., Elisseeff, J.H. & Yu, S.M. (2006). Collagen Mimetic Peptide-Conjugated Photopolymerizable PEG Hydrogel. *Biomaterials*, Vol.27, No.30, (October 2006), pp. 5268-5276, ISSN 0142-9612

Lee, H.J., Yu, C., Chansakul, T., Hwang, N.S., Varghese, S., Yu, S.M. & Elisseeff, J.H. (2008). Enhanced Chondrogenesis of Mesenchymal Stem Cells in Collagen Mimetic Peptide-Mediated Microenvironment. *Tissue Engineering Part A*, Vol.14, No.11, (November 2008), pp. 1843-1851, ISSN 1076-3279

Lee, S.H. & Shin, H. (2007). Matrices and Scaffolds for Delivery of Bioactive Molecules in Bone and Cartilage Tissue Engineering. *Advanced Drug Delivery Review*, Vol.59, No.4-5, (May 2007), pp. 339-359, ISSN 0169-409X

Liu, Y., Shu, X.Z. & Prestwich, G.D. (2006). Osteochondral Defect Repair with Autologous Bone Marrow-Derived Mesenchymal Stem Cells in an Injectable, in Situ, Cross-Linked Synthetic Extracellular Matrix. *Tissue engineering*, Vol.12, No.12, (December 2006), pp. 3405-3416, ISSN 1076-3279

Lutolf, M.P., Raeber, G.P., Zisch, A.H., Tirelli, N. & Hubbell, J.A. (2003). Cell-Responsive Synthetic Hydrogels. *Advanced Materials*, Vol.15, No.11, (June 2003), pp. 888-892, ISSN 0935-9648

Martens, P.J., Bryant, S.J. & Anseth, K.S. (2003). Tailoring the Degradation of Hydrogels Formed from Multivinyl Poly(ethylene glycol) and Poly(vinyl alcohol) Macromers for Cartilage Tissue Engineering. *Biomacromolecules,* Vol.4, No.2, (March-April 2003), pp. 283-292, ISSN 1525-7797

Mather, B.D., Viswanathan, K., Miller, K.M. & Long, T.E. (2006). Michael Addition Reactions in Macromolecular Design for Emerging Technologies. *Progress in Polymer Science,* Vol.31, No.5, (May 2006), pp. 487-531, ISSN 0079-6700

McHale, M.K., Setton, L.A. & Chilkoti, A. (2005). Synthesis and in Vitro Evaluation of Enzymatically Cross-Linked Elastin-Like Polypeptide Gels for Cartilaginous Tissue Repair. *Tissue Engingeering,* Vol.11, No.11-12, (January 2005), pp. 1768-1779, ISSN 1076-3279

Nettles, D.L., Kitaoka, K., Hanson, N.A., Flahiff, C.M., Mata, B.A., Hsu, E.W., Chilkoti, A. & Setton, L.A. (2008). In Situ Crosslinking Elastin-Like Polypeptide Gels for Application to Articular Cartilage Repair in a Goat Osteochondral Defect Model. *Tissue Engineering Part A,* Vol.14, No.7, (July 2008), pp. 1133-1140, ISSN 1076-3279

Nguyen, K.T. & West, J.L. (2002). Photopolymerizable Hydrogels for Tissue Engineering Applications. *Biomaterials,* Vol.23, No.22, (November 2002), pp. 4307-4314, ISSN 0142-9612

Obara, K., Ishihara, M., Ishizuka, T., Fujita, M., Ozeki, Y., Maehara, T., Saito, Y., Yura, H., Matsui, T., Hattori, H., Kikuchi, M. & Kurita, A. (2003). Photocrosslinkable Chitosan Hydrogel Containing Fibroblast Growth Factor-2 Stimulates Wound Healing in Healing-Impaired db/db Mice. *Biomaterials,* Vol.24, No.20, (September 2003), pp. 3437-3444, ISSN 0142-9612

Ono, K., Saito, Y., Yura, H., Ishikawa, K., Kurita, A., Akaike, T. & Ishihara, M. (2000). Photocrosslinkable Chitosan as a Biological Adhesive. *Journal of Biomedical Mterials Research,* Vol.49, No.2, (February 2000), pp. 289-295, ISSN 1549-3296

Park, H., Temenoff, J.S., Holland, T.A., Tabata, Y. & Mikos, A.G. (2005). Delivery of TGF-Beta1 and Chondrocytes Via Injectable, Biodegradable Hydrogels for Cartilage Tissue Engineering Applications. *Biomaterials,* Vol.26, No.34, (December 2005), pp. 7095-7103, ISSN 0142-9612

Park, K.H. & Na, K. (2008). Effect of Growth Factors on Chondrogenic Differentiation of Rabbit Mesenchymal Cells Embedded in Injectable Hydrogels. *Journal of Bioscience and Bioengineering,* Vol.106, No.1, (July 2008), pp. 74-79, ISSN 1389-1723

Park, Y., Sugimoto, M., Watrin, A., Chiquet, M. & Hunziker, E.B. (2005). BMP-2 Induces the Expression of Chondrocyte-Specific Genes in Bovine Synovium-Derived Progenitor Cells Cultured in Three-Dimensional Alginate Hydrogel. *Osteoarthritis Cartilage,* Vol.13, No.6, (June 2005), pp. 527-536, ISSN 1063-4584

Park, Y.D., Tirelli, N. & Hubbell, J.A. (2003). Photopolymerized Hyaluronic Acid-Based Hydrogels and Interpenetrating Networks. *Biomaterials,* Vol.24, No.10, (March 2003), pp. 893-900, ISSN 0142-9612

Park, Y.D., Lutolf, M.P., Hubbell, J.A., Hunziker, E.B. & Wong, M. (2004). Bovine Primary Chondrocyte Culture in Synthetic Matrix Metalloproteinase-Sensitive Poly(ethylene glycol)-Based Hydrogels as a Scaffold for Cartilage Repair. *Tissue engineering,* Vol.10, No.3-4, (March-April 2004), pp. 515-522, ISSN 1076-3279

Sanborn, T.J., Messersmith, P.B. & Barron, A.E. (2002). In Situ Crosslinking of a Biomimetic Peptide-PEG Hydrogel Via Thermally Triggered Activation of Factor XIII *Biomaterials,* Vol.23, No.13, (July 2002), pp. 2703-2710, ISSN 0142-9612

Seliktar, D., Zisch, A.H., Lutolf, M.P., Wrana, J.L. & Hubbell, J.A. (2004). Mmp-2 Sensitive, Vegf-Bearing Bioactive Hydrogels for Promotion of Vascular Healing. *Journal of Biomedical Materials Research: Part A,* Vol.68, No.4, (March 2004), pp. 704-716, ISSN 1549-3296

Silver, D., Kapsch, D.N. & Tsoi, E.K. (1983). Heparin-Induced Thrombocytopenia, Thrombosis, and Hemorrhage. *Annuals of Surgery,* Vol.198, No.3, (September 1983), pp. 301-306, ISSN 0003-4932

Skjak-Brvk, G., Grasdalen, H. & Smidsrod, O. (1989). Inhomogeneous Polysaccharide Ionic Gels. *Carbohydrate Polymer,* Vol.10, No.1, (February 1989), pp. 31-54, ISSN 0144-8617

Sontjens, S.H.M., Nettles, D.L., Carnahan, M.A., Setton, L.A. & Grinstaff, M.W. (2006). Biodendrimer-Based Hydrogel Scaffolds for Cartilage Tissue Repair. *Biomacromolecules,* Vol.7, No.1, (January 2006), pp. 310-316, ISSN 1525-7797

Sperinde, J.J. & Griffith, L.G. (1997). Synthesis and Characterization of Enzymatically-Cross-Linked Poly(Ethylene Glycol) Hydrogels. *Macromolecules,* Vol.30, No.(September 1997), pp. 5255-5264, ISSN 0024-9297

Tomme, S.R.v., Steenbergen, M.J.v., Smedt, S.C.d., Nostrum, C.F.v. & Hennink, W.E. (2005). Self-Gelling Hydrogels Based on Oppositely Charged Dextran Microspheres. *Biomaterials,* Vol.26, No.14, (May 2005), pp. 2129-2135, ISSN 0142-9612

Wang, C., Stewart, R.J. & KopeCek, J. (1999). Hybrid Hydrogels Assembled from Synthetic Polymers and Coiled-Coil Protein Domains. *Nature,* Vol.397, No.6718, (February 1999), pp. 417-420, ISSN 0028-0836

Wang, D.A., Williams, C.G., Yang, F. & Elisseeff, J.H. (2004). Enhancing the Tissue-Biomaterial Interface: Tissue-Initiated Integration of Biomaterials. *Advanced Functional Materials,* Vol.14, No.12, (December 2004), pp. 1152-1159, ISSN 1616-3028

Wang, D.A., Varghese, S., Sharma, B., Strehin, I., Fermanian, S., Gorham, J., Fairbrother, D.H., Cascio, B. & Elisseeff, J.H. (2007). Multifunctional Chondroitin Sulphate for Cartilage Tissue-Biomaterial Integration. *Nature Materials,* Vol.6, No.5, (May 2007), pp. 385-392, ISSN 1476-1122

Wathier, M., Jung, P.J., Carnahan, M.A., Kim, T. & Grinstaff, M.W. (2004). Dendritic Macromers as in Situ Polymerizing Biomaterials for Securing Cataract Incisions. *Journal of the American Chemical Society,* Vol.126, No.40, (September 2004), pp. 12744-12745, ISSN 0002-7863

Wathier, M., Johnson, C.S., Kim, T. & Grinstaff, M.W. (2006). Hydrogels Formed by Multiple Peptide Ligation Reactions to Fasten Corneal Transplants. *Bioconjugated Chemistry,* Vol.17, No.4, (June 2006), pp. 873-876, ISSN 1043-1802

Weng, L., Gouldstone, A., Wu, Y. & Chen, W. (2008). Mechanically Strong Double Network Photocrosslinked Hydrogels from N, N-Dimethylacrylamide and Glycidyl Methacrylated Hyaluronan. *Biomaterials,* Vol.29, No.14, (February 2008), pp. 2153-2163, ISSN 0142-9612

Wichterle, O. & Lim, D. (1960). Hydrophilic Gels in Biologic Use. *Nature,* Vol.185, No.1, (January 1960), pp. 117-118, ISSN 0028-0836

Xu, C. & Kopeček, J. (2008). Genetically Engineered Block Copolymers: Influence of the Length and Structure of the Coiled-Coil Blocks on Hydrogel Self-Assembly. *Pharmaceutical Research,* Vol.25, No.3, (March 2008), pp. 674-682, ISSN 0724-8741

Yang, J., Xu, C., Kopeckova, P. & Kopecek, J. (2006a). Hybrid Hydrogels Self-Assembled from HPMA Copolymers Containing Peptide Grafts. *Macromolecular Bioscience,* Vol.6, No.3, (March 2006a), pp. 201-209, ISSN 1616-5195

Yang, J., Xu, C., Wang, C. & Kopecek, J. (2006b). Refolding Hydrogels Self-Assembled from N-(2-Hydroxypropyl)Methacrylamide Graft Copolymers by Antiparallel Coiled-Coil Formation. *Biomacromolecules,* Vol.7, No.4, (April 2006b), pp. 1187-1195, ISSN 1525-7797

Yang, S., Leong, K.-F., Du, Z. & Chua, C.-K. (2001). The Design of Scaffolds for Use in Tissue Engineering. Part I. Traditional Factors. *Tissue engineering,* Vol.7, No.6, (December 2001), pp. 679-689, ISSN 1076-3279

Yayon, A., Klagsbrun, M., Esko, J.D., Leder, P. & Ornitz, D.M. (1991). Cell Surface, Heparin-Like Molecules Are Required for Binding of Basic Fibroblast Growth Factor to Its High Affinity Receptor. *Cell,* Vol.64, No.4, (February 1991), pp. 841-848, ISSN 0092-8674

Yu, L., Zhang, H. & Ding, J. (2006). A Subtle End-Group Effect on Macroscopic Physical Gelation of Triblock Copolymer Aqueous Solutions. *Angewandte ChemieInternational Edition,* Vol.45, No.14, (March 2006), pp. 2232-2235, ISSN 1433-7851

Yu, Y.B. (2002). Coiled-Coils: Stability, Specificity, and Drug Delivery Potential. *Advanced Drug Delivery Review,* Vol.54, No.8, (October 2002), pp. 1113-1129, ISSN 0169-409X

Yuji, H., Rocky, S.T. & Lillian, S. (2004). Distinct Functions of BMP4 and GDF5 in the Regulation of Chondrogenesis. *Journal of Cell Biochemistry,* Vol.91, No.6, (April 2004), pp. 1204-1217, ISSN 1097-4644

Part 2

Gene Medicine and Nanobiomedicine

3

RNA Interference for Tumor Therapy

Wei Xia and Jing Ni
Department of Nuclear Medicine, Shanghai Seventh People's Hospital, Shanghai, P.R. China

1. Introduction

The most recent investigation from the World Health Organization's International Agency for Research on Cancer (WHO/IARC) indicates that the incidence of malignant tumors is rising throughout the world. In 2008, 7.6 million people died of malignant tumors and 64% of them occurred in developing countries(Jemal *et al,* 2011). Particularly, the status on the treatment of tumors is extremely harsh in China since both incidence and mortality of malignant tumors in China is drastically rising in the past two decades (Zhang *et al,* 2008). It is expected that this increasing tendency will continue in China in the next two decades(Xv&Dong,2003). There is no doubt that tumor therapy will be for long term a hot topic in the field of life and medicine in the feature.

The extensive understanding of molecular mechanisms underlying tumor development and progression provides more deep insights that tumor is a genetic disease. It appears that the correction of defective genes would likely be a new approach for the treatment of tumors (Wang *et al,* 2010). RNA interference (RNAi) is a double-stranded RNA-induced gene silencing process, by which RNA can specifically silence complementary mRNA in the cells. Fire *et al.* for the first time discovered this phenomenon in the cells of rod-shaped nematomorpha(Fire *et al,* 1998), for which he was awarded the Nobel Prize in Physiology or Medicine in 2006. Further studies found that RNAi exists in almost all multi-cellular organisms and plays an important role in defense against double-stranded RNA virus(Hannon,2002). Also, RNAi-based tumor therapy in animal models has demonstrated that RNAi is a powerful means for tumor therapy(López-Fraga *et al,* 2009;Davis *et al,* 2010; Koldehoff& Elmaagacli,2009; Maples *et al,* 2009).

This chapter aims to review current progress in RNAi for tumor therapy. The mechanisms underlying RNAi-induced gene silencing pathway were briefly described in the first section. Next, the application of RNAi targeting different tumor-related genes was summarized in detail. Finally, recent achievement of RNAi-based tumor therapy in clinical trials was also reviewed.

2. Mechanisms of RNAi

The RNAi phenomenon was first found in antisense hybridization of antisense RNA and mRNA. However, further studies showed that the sense strand could also result in reduced gene expression (Guo& Kemphues,1995). Subsequent studies demonstrated that each strand of dsRNA can induce RNAi, but only few types of dsRNA can lead to potent interference effect (Fire *et al,* 1998). The molecules that can trigger RNAi include small interference RNA (siRNA), microRNA (miRNA) and small hairpin RNA (shRNA). The mechanisms of RNAi triggered by these molecules are not exactly the same.

2.1 Mechanisms of siRNA

SiRNA is a class of dsRNA molecules with 20-25 nucleotides in length. When dsRNA enters cells, it can be lysed into siRNA by Dicer enzyme. On the other hand, dsRNA proliferates in the presence of RNA-directed RNA polymerase (RdRP) and is then lysed in the presence of Dicer enzyme. Double-stranded siRNA is processed by Dicer into single-stranded siRNA, which can form a complex via binding with some related proteins. This complex can bind simultaneously with complementary mRNA, thereby inducing mRNA lysis.

SiRNA-induced RNAi is initiated from a cleavage program. Long dsRNA is first cleaved by Dicer into siRNAs in cytoplasm. The Dicer contains an N-terminal helicase domain, a PAZ (Piwi/Argonaute/Zwille) RNA domain, two RNAase III domains, and a dsRNA binding domain (dsRBD) (Bernstein *et al,* 2001). The RNA helicase domain sequence is highly stable and N-terminal and C-terminal are core components. The N-terminal has an ATP-binding and hydrolysis subunit, and the C-terminal contains a RNA-binding subunit. With these subunits, the RNA helicase domain can play a helicase function by making use of the energy provided by ATP (Ma *et al,* 2004). The PAZ domain of Dicer can specifically recognize and bind with the 5′ phosphate bond with two nucleotide protrusions in the dsRNA3′ end (Cordin *et al,* 2006). There is a long α-helix between the PAZ domain and the RNase III domain, and this helix is encircled by a conserved N-terminal protein, forming a platform domain consisting of an anti-parallel β-sheet and three α-helixes, and at both ends of the platform domain are hinge1 and hinge 2 as a polypeptide chain to hinge with the PAZ domain and the RNase III domain respectively. Of them, hinge1 contains a Pro-266 kink, which enables the PAZ domain to move a 5Å distance. Unlike hinge 1, hinge 2 does not have the link, but it can enable the RNase III domain to make a 5° rotation along the long axis of the Dicer molecule. Therefore, Dicer is able to adapt to substrates of different shapes by regulating the platform, hinge 1 and hinge 2 for the purpose of cleaving different substrates (Macrae *et al,* 2006). Each RNase III domain possesses two active centers, and each center has the activity of cleaving the phosphodiester bond, and can cleave dsRNA into small RNA with two nucleotide protrusions in the dsRNA3′ end, which is the typical characteristic of the RNase III product (Collins&Cheng,2005). Although Dicer contains two RNase III domains and four active cleavage centers, each RNase III domain provides only one active center for cleavage. The distance between the active centers of the two RNase III domains is 65Å apart, equaling 25 bp RNA in length. This structural characteristic ensures Dicer to specifically produce small RNA of 21~25 bp in length, and enables it to change the length of the cleaved product by mirco-regulating the distance between the two RNase III domains (Blaszczyk *et al,* 2001). In the C-terminal of Dicer is the dsRBD that contains about 70 amino acids existing in the form of αβββα, of which the α-helix is close to one side of β-sheet in a reverse and parallel form (Hallegger *et al,* 2006). dsRBD and dsRNA interact with each other across a width of 16 bp, which is just the distance between the two small adjacent grooves and the large groove on one side of α-dsRNA(Wu *et al,* 2004). This interacting region can be further divided into three small regions. Region 1: the four 2′-OH at the first small groove of dsRBD N-terminal α-helix connect with five nucleotides of RNA. Region 2: a stem loop between β1and β2 of dsRBD connects with 2-5 nucleotides of RNA in the second small groove through 2′-OH. Region 3: the α-helix at the C-terminal of dsRBD travels across the large groove of RNA and connects with the phosphodiester bond in RNA (Ryter & Schultz, 1998).

The long dsRNA that enters cells is processed by Dicer into double-stranded siRNA, and then bind with Dicer, trans-activitor response region RNA-binding protein (TRBP), and protein kinase R (PKR)-activating protein (PACT) to form a RNA-induced silencing complex (RISC) (Hammond *et al,* 2000).

RNAi induced by completely processed single-stranded siRNA (mostly exogenous) can be directly installed into RISC without the participation of TRBP/PACT-Dicer complex. However, in other RNAi processes, RNAi cannot be installed into RISC without the participation of TRBP/PACT-Dicer complex. Through TRBP/PACT, siRNA-TRBP/PACT-Dicer complex binds with Argonaute 2 (Ago2/EIF2C2) to form RISC loading complex (RLC) (Robb&Rana,2007). RLC initiates disentanglement of siRNA and distinguishs the guide strand and the passenger strand. The guide strand 5′-end has the property of a low thermodynamic stability and can be bound into RISC by priority(Tomari *et al,* 2004). The passenger strand is cleaved by Ago 2 and separated from RLC (Matranga *et al,* 2005). The specific sequence for binding between the RISC-mediated RNA guide strand and target mRNA is mainly located in the "seed" or "core" region of 2-8 nucleotides of siRNA 5′-end (Ma *et al,* 2005). mRNA oligonucleotide is cleaved from the middle in the presence of Ago2 and loses the expression function (Schwarz *et al,* 2004; Hammond,2005).

Ago 2 is the only protein that has the endonuclease function in RLC, mainly consisting of an N-terminal domain, a PAZ domain, a middle domain and a PIWI domain, of which the N-terminal domain, middle domain and PIWI domain form a crescent basin structure, with one "arm" supporting the PAZ domain above the "basin" (Song *et al,* 2004;Yuan *et al,* 2005). This structure forms a groove carrying positive charge. The N-terminal domain and PIWI domain form the small groove, which can bind with nucleic acid. The PAZ domain and the crescent "basin" form the large groove, which can contain double strands and locate the easily broken phosphate bond directly to the catalytic site of Rnase H sheet on the PIWI domain. Ago 2 can melt siRNA, forming single-stranded antisense siRNA complementary to target mRNA. The Rnase H PIWI domain of Ago 2 is located in the 5′-end, and the PAZ domain can recognize siRNA3′-end. When siRNA binds with complementary target mRNA, Ago 2 begins digesting mRNA, forming RNA fragments with 5′-carboxylate and 3′-phosphate to degrade mRNA of the endogenous target gene, thus preventing gene expression. During the process of degradation, mRNA comes in between the N-terminal and the PAZ domain and goes out between the PAZ domain and the middle domain. When mRNA moves between the PAZ domain and the "crescent basin", it is cleaved by the active site of the PIWI domain.

RISC can be recycled. When a target mRNA is degraded, RISC can continue to degrade other target mRNA, thus amplifying the PTGS effect of the target gene (Gregory *et al,* 2005) and this amplification can maintain for few days in dividing cells. Obviously, it is advantageous over other therapeutic methods, and is a potent method for specific inhibition of gene expression. In addition, a few studies(Meister *et al,* 2005) also found that RISC also has some other protein components, including Vasa intronic gene (VIG) protein, DmFXR, Tudor-SN, potential RNA helicase Dmp68 and Gemin3. These proteins are not necessary for the activity of RISC nuclease, but may provide additional functions such as RISC turnover, RISC subcellular location, and degradation after mRNA cleavage.

2.2 Mechanisms of miRNA

miRNAs are a class of endogenous non-coding RNA that serves a specific function in eukaryotic organisms, with about 20~25 nucleotides. Mature miRNAs are processed from relatively long primary transcripts through a series of nuclease cutting, and then assembled into the RNA-induced silencing complex to identify target mRNA through base pairing, and instruct the silencing complex to degrade or block the translation of target mRNA depending on the degree of complementation.

miRNA genes are usually transcribed by RNA polymerase II (pol II) in the nucleus, and the primary product is pre-miRNA with a large cap-structure (7MGpppG) and a poly-A tail (AAAAA). Pre-miRNA is treated in stem loop-structured pre-mRNA containing 60-110 nucleotides in the presence of Rnase III enzyme Drosha (RNASEN) and its cofactor DGCR8 (a double-strand RNA domain-binding protein)(Tang,2005), and this pre-miRNA is transported from the nucleus to the cytoplasm by RAN-GTP and exportin 5 (EXP 5), and then cleaved by another nuclease Dicer to produce miRNA:miRNA double strands about 22 nucleotides in length, which are quickly introduced into RISC containing Ago2. One of the single-stranded mature miRNA is retained in the complex and expressed in response to regulatory genes in a manner similar to siRNA at the compensatory site of mRNA via base pairing (Meister&Tuschl,2004). Apart from Ago2, Ago1, Ago3 and Ago4 in the argonaute family that have no endonuclease activity also participate in the RNAi process of RISC through regulation on RISC rather than through cleavage (Preall&Sontheimer,2005).

Identification of target mRNA by miRNA requires precise pairing of nucleotides with miRNA seed sequence (Sethupathy *et al,* 2006). If miRNA and mRNA compensate with each other precisely, Ago 2 can cleave mRNA in the same way as it cleaves siRNA (Meister *et al,* 2004). However, there is a convex sequence in the middle of the A-type helical structure from non-precise combination of compensatory miRNA and the target mRNA, which makes mRNA cleavage impossible but can inhibit the expression of target mRNA at the protein translation level. This cleavage-independent mechanism plays the main role in miRNA-nduced RNAi (Pillai *et al,* 2007). The miRNA-binding site using this mechanism is usually in 3′-end non-translation region of mRNA. Each miRNA can have multiple target genes, while several miRNAs can also regulate the same gene. This complex regulatory network can regulate expression of multiple genes through one miRNA, or precisely regulate the expression of a particular gene through combination of several miRNAs (Sethupathy *et al,* 2006). About one-third of encoding protein mRNA can be regulated by miRNA(Krek *et al,* 2005).

2.3 Mechanisms of shRNA

shRNA contains two short inverted repeat sequences separated by a stem loop in the middle forming a hairpin structure, which is controlled by Pol III promoter and then connects with 5-6 T as the transcription terminator of RNA Pol III. One of the methods of transporting siRNA in vivo is to clone the siRNA sequence into the plasmid as a "short hairpin". When the hairpin is delivered to the animal body, its sequence is expressed, forming "double-stranded RNA" (shRNA), which is treated by the RNAi pathway.

The first-generation shRNA is decorated and cloned to the viral carrier on the pre-miRNA platform, and transcribed into 50-70bp single-stranded molecules with the unique "stem-loop" structure in the presence of RNA Pol III in the nucleus. When it enters the cytoplasm, its stem-loop structure is cut off by Dicer, forming siRNAs. Finally, it participates in the production of the RISC complex and induces RNAi (Paddison *et al,* 2004; Brummelkamp *et al,* 2003; Berns *et al,* 2004). With further research on miRNA, researchers have re-modeled the well-known structure of human miRNA-mir-30 and successfully developed a carrier-mediated second-generation RNAi trigger shRNAmir using pri-miRNA as the platform. Unlike pri-mir-30, the "stem" of this carrier-mediated shRNAmir has been replaced by other double helixes specific to different target genes. This structural re-design and transformation do not affect normal maturity of mir-30. Compared with the first-generation shRNA, the design of the second-generation shRNAmir makes use of the endogenous Drosha digestion process, thus increasing the specificity of Dicer in identifying Pre-miRNA. In addition, shRNAmir enters the intracellular RNAi pathway than shRNA or siRNA and can be treated by Drosha and Dicer, thus producing more siRNA(Silva *et al,* 2005).

2.4 Mechanisms of bifunctional shRNA

Bifunctional shRNA can make use of cleavage-dependent and non cleavage-dependent RNAi (Rao *et al,* 2010). Such shRNA contains two stem loops: one with complete-matching passenger and guide strands, which can combine with cleavage-dependent RISC, and the other with non-matching strands, which can combine with non cleavage-dependent RISC. This shRNA is more efficient than shRNA RNAi with one RISC complex. In experiments directly comparing the same siRNA target RNAi, it was found that bifunctional shRNA was more efficient (the same dosage of shRNA could produce more powerful gene knockout function) and its pharmacokinetics also underwent change(Rao *et al,* 2009,2010).

3. The application of RNAi for silencing of tumor-related genes

RNAi has a standard procedure in intracellular gene knockout and can be used to identify gene functions by inhibiting specific genes and analyzing the change of gene appearance. As such, by analyzing gene functions, the signal pathway of tumor development could be identified. This further provides the feasibility to assess the medical or RNAi therapeutic target(Devi,2006). In addition, systemic treatment approaches with cancer-related miRNA or miRNA offer opportunities for tumor therapy. There have been a few overexpressed target genes for RNAi study *in vivo*. The section is a retrospective review of these targets.

3.1 Vascular endothelial growth factor

Vascular endothelial growth factor (VEGF) is a heparin-binding growth factor specific to vascular endothelial cells, and can induce angiogenesis in vivo. VEGF and anti-VEGF coexist in normal tissues and remain in a relatively balanced state. In the presence of tumor growth, multiple carcinogenic factors trigger dramatic increase in VEGF, the amount and function of which are by far greater than those of anti-VEGF, thus causing massive growth of vessels (mainly blood vessels) and promoting tumor growth and metastasis. In addition, VEGF interferes with and inhibits dendritic cells and blocks antigen presentation of B and T cells, which further induces immune escape of tumors, interferes with the normal immune function of the body, and causes resistance to anti-tumor therapy. He (He *et al,* 2009) used $CaCO_3$ as the matrix and form nanoscaled complexed with the plasmid to expresses siRNA targeting VEGF. In their siRNA transfection against colon cancer cells in vitro, they found that this complex could significantly reduce the expression level of VEGF-C in LoVo cells. In the experiment with transplanted colon cancer in nude mice, the mean tumor size treated by the complexes was decreased by 50% as compared to that in the control group. Metastasis of the cancer cells in lymph nodes and lymphatic ducts was also inhibited.

3.2 Vimentin

Cell structures and morphologies are maintained through a series of complex structural proteins, including microtubules, microfilaments and intermediate filaments. Vimentin is a type III intermediate filament. Studies in recent years have demonstrated that vimentin is closely associated with tumor development and metastasis, and participates in adhesion, migration, invasion and cell signal transduction of tumor cells and tumor-related endothelial cells and macrophages. Vimentin possesses a highly dynamic balance between polymerization depolymerization and a very complex phosphorylated form, which serve as the basis for it to participate in tumor metastasis and intercellular actions. Vimentin is found to be homological to most sequences of miR-17-3p in the miRNA gene cluster 17-92 family. miR-17-3p participates in regulating the expression of vimentin. miR-17-3p is a known tumor-inhibiting gene. It remains at a low level in highly tumorigenic and metastatic cell lines and at a high level in cell lines of low tumorigenesis. Zhang (Zhang *et al,* 2009) used plasmds expressing miR-17-3p to intervene transplanted prostate cancer in mice and found that the tumor size was only 50% of the control group 31 days after the experiment. They also found that vimentin expression was negatively correlated with miR-17-3p expression, and the rate of tumor growth was also negatively correlated with miR-17-3p. Paccione RJ (Paccione *et al,* 2008)used vimentin as the target to perform RNA interference3, and found that the ability of cell proliferation, metastasis and invasion in cellRNAi group decreased by 3 fold as compared with the control group. In addition, tumorigenesis of cells treated with vimentin-RNAi decreased significantly, and the size of tumors injected subcutaneously with vimentin-RNAi was 70% smaller than that of the control group.

3.3 STAT3

Signal transducer and activator of transcription 3 (STAT3) is a member of the STAT family, participating in signaling pathways closely related to cell proliferation, differentiation and apoptosis. Its abnormal activity may promote cell differentiation and proliferation and inhibit cell apoptosis, causing tumor development and promoting metastasis of tumor cells. It is found to be activated persistently in hepatocellular carcinoma (HCC), breast cancer and esophageal squamous cell carcinoma. Li (Li *et al,* 2009) injected STAT3-target shRNA, negative shRNA and normal saline (NS) into transplanted HCC tumors in nude mice and removed the tumors after observation for a period. It was found that the tumor tissue in STAT3-shRNA, negative shRNA and NS groups was 0.18g, 0.6g and 0.67g respectively. It was also found that the expression of STAT3, phosphorylation of STAT3, and expressions of

VEGF, survivin and MYC were down-regulated, while caspase3 and p53 were up-regulated in STAT3-shRNA group. Huang (Huang *et al,* 2011)studied the effect of RNAi using STAT3 as the target gene on pancreatic cancer, and found that STAT3 RNAi not only inhibited tumor growth but inhibited angiogenesis in the tumors. They also found that VEGF and MMP-2 mRNA and protein levels decreased significantly in SW1990 cells treated with RNAi, indicating that STAT3 plays an important role in the course of tumor development and progression, and can be used as a target for tumor gene therapy.

3.4 Human telomerase reverse transcriptase

Human telomerase reverse transcriptase (hTERT) is a rate-limiting enzyme maintaining the activity of telomerase. However, it is found that hTERT not only maintains telomerase but stabilizes cell genomes, promotes cell proliferation, inhibits cell apoptosis, and plays a role in regulating intracellular signaling pathways. A recent study(de Souza Nascimento *et al,* 2006)transfected Hela with shRNA expressing carrier-encoded target shRNA, and found that short telomerase-related hTERT mRNA was reduced dramatically, cell growth was inhibited, and cell apoptosis was accelerated. A similar study(Gandellini *et al,* 2007)showed that transfection of hTERT specific siRNA with lipofectamine 2000 could inhibit the expression of hTERT for a prolonged time, thus affecting the proliferation and tumorigenesis of prostate cancer PC-3 cells.

3.5 Survivin

Survivin is a recently discovered tumor-specific inhibitor of the apoptosis protein family. There is no or micro expression in well differentiated tissues, and high expression in embryonic and most human malignant tumor tissues. It inhibits apoptosis of tumor cells, thus promoting cell growth and regulating cell division. In addition, it is closely associated with infiltration and metastasis of tumor cells. Survivin is found expressing in common malignant tumors such as breast cancer, gastric cancer, kidney cancer, melanoma, intestinal cancer, neuroblastoma cancer and ovarian cancer. In their retrospective analysis of tumor treatment, Kanwar (Kanwar *et al,* 2010)demonstrated that survivin-targeted molecular therapy could inhibit the growth of tumor cells. Recently, Chen (Chen&Deng,2008) injected gastric cancer MGC-803 cells treated with plasmids expressing survivin-siRNA and control-treated plasmids, and those without intervention into nude mice subcutaneously. Four weeks later, the mean tumor volume in the intervention group, control group and non-intervention group was 831mm^3, 2617mm^3 and 2536mm^3 respectively, indicating that the apoptosis rate in siRNA intervention group was significantly higher than that in the control and non-intervention groups (27.63% vs 2.15% vs 2.31%). In their experiment study with survivin-target RNAi for the treatment of urinary bladder cancer, Seth (Seth *et al,* 2011) reported the similar result and further confirmed that survivin-target RNAi could inhibit tumor cell growth effectively.

3.6 ERK-MAPK

ERK-MAPK is activated in many malignant tumors, and plays a primarily important role in cell growth, differentiation and survival. siRNA specific to extracelluar signal regulating kinase 1/2 (ERK1/2) could inhibit the proliferation of ovarian cancer cells and induce their apoptosis and death, but it had no effect on normal ovarian cells (Zeng *et al,* 2005). Super-expression of epithelial growth factor receptor HER2/neu was detected in many breast cancers, but its expression in patients with good prognosis was low. Transfection of HER2/neu siRNA led to an antiproliferative and apoptosis-inducing effect in breast cancer cells with high expression of HER2, and this effect was not observed in cells without HER2 expression, suggesting that interference with HER2 could be used as an effective strategy for the treatment of breast cancer with high expression of HER2(Faltus *et al,* 2004).

3.7 Osteopontin

Osteopontin (OPN) is a secretory extracellular matrix protein and a cytokine as well. It can promote cell chemotaxis, adhesion and migration. More studies (Wai&Kou, 2008; Wise & King, 2008;Ramaiah&Rittling,2008) have demonstrated that OPN can interact with $\alpha V\beta 3$ and CD44 to participate in cell inflammatory response, promote cell migration, adhesion and transduction of related signals, regulate the expression of genes related to tumor development and progression, promote degradation of extracellular matrix, suppress the immune function of the body against tumors, and accelerate tumor progression. Studies(Coppola *et al,* 2004; Cui *et al,* 2007;Tang *et al,* 2007)have found that the level of OPN expression is generally increased in lung, gastrointestinal, urogenital and gynecological tumors. Zhang (Zhang *et al,* 2010) used plasmid-mediated OPN-target siRNA to transfect lung cancer A549 cells and found that the invasion and proliferation abilities of A549 cell decreased. Gong (Gong *et al,* 2010) used PEI as the carrier of OPN-siRNA and injected it between tumor tissues in nude mice bearing transplanted gastric cancer tumors. It was found that the tumor size of the experiment group was only one-third of the control group, and survival of the mice was also prolonged (animals in the control group survived 40 days, and 50% of the animals in the experiment group remained surviving at 60 days).

3.8 β-catenin

β-catenin is a multi-functional adapter protein encoded by human CTNNB1 gene. Its function is expressed in two aspects: One is acting as an important component of the Wnt signaling pathway to mediate signal movement from the membrane to the cytoplasm and to the nucleus, regulating transcription of target genes (such as c-myc,cyclin D1, MMP-7 and cyclooxygenase-2). The other is participating in E-cadherin-mediated intercellular adhesion. β-catenin participates in cell growth, proliferation, invasion and metastasis in malignant tumor tissues. It was found that β-catenin expression was extraordinarily high in many tumors, especially in esophageal cancer. Wang (Wang *et al,* 2009,2010) used β-catenin-targeted RNA to interfere with plasmid pGen-3-CTNNB1 and found that pGen-3-CTNNB1 could lower the expression level of β-catenin in esophageal cancer tissue, and reduce the invasion and metastasis abilities of cancer cells. The result of the experiment with nude mice bearing transplanted esophageal cancer showed that the mean volume of tumors in β-catenin RNAi plasmid group was $909.3mm^3$ versus $2684.4mm^3$ in the negative control group and $2722.6mm^3$ in the non-intervention group, indicating that β-catenin-target RNAi could inhibit the growth of esophageal cancer tumors.

3.9 MYC

Myc gene is a group of cancer genes discovered earlier, including C-myc, N-myc and L-myc, belonging to nucleoprotein-coding cancer gene. All the three genes encode a nuclear DNA-binding protein related to cell cycle regulation. The Myc gene family and its products can promote cell proliferation, immortalization, dedifferentiation and transformation, and play an important role in the formation of various tumors. Super-expression of Myc gene has been detected in gastric cancer, breast cancer, colon cancer, cervical cancer, Hodgkin's disease and head tumors. Zhang (Zhang *et al,* 2009)injected Myc-target siRNA between transplanted colon cancer tissues in nude mice, and found that the expression level of Myc decreased by 40%; large patches of necrosed cells were seen in tumor tissues; the proliferation of tumor cells was inhibited; and the mean tumor size was about 50% of the control group.

3.10 Epithelial cellular adhesion molecule

Epithelial cellular adhesion molecule (EpCAM) is a recognized tumor-related protein encoded by TACSTDl gene. EpCAM is more than an adhesion molecule, and also has many other functions including participation in regulating cell migration, metastasis, differentiation, signal transduction, cell cycle and metabolism. Studies have demonstrated that EpCAM presents high expression in most solid tumors. In addition, the level of EpCAM expression is positively correlated with lymphatic metastasis. Du (Du *et al,* 2009) transfected lipofectamine 2000 with EpCAM-target siRNA to treat two gastric cell lines (AGS and SGC7901) in transplanted tumor-bearing nude mice, and found that the size of both tumors was about 50% smaller than that of the control group. They also found that the level of CCND1 expression in tumor tissues was decreased, and cell division was blocked at G1.

3.11 CD147

CD147 is a transmembrane glycoprotein extensively expressing in various tissues of the human body, belonging to the immunoglobulin super-family participating in various physilogical and pathological processes of the body. CD147 interacts with integrin, central avidin, monocarboxylate transporter 1 (MCT1), MCT3 and MCT4 proteins. It is also related to the expression of lactate transporters. These proteins may act as candidate ligands or receptors of CD147 molecule and mediate the biological function of epithelial cells via protein-protein interactions. High expression of CD147 is detected in multiple tumor tissues, especially in pancreatic cancer (Zhang *et al,* 2007), where it promotes invasion, metastasis and anchorage-independent growth of tumor cells, and induces tumor angiogenesis. Schneiderhan (Schneiderhan *et al,* 2009) used CD147-target RNAi to treat pancreatic cancer cells and found that MCT1 and MCT4 were inhibited in tumor cells, and the intracellular concentration of lactate was increased, thus inhibiting the growth of tumor cells. In vivo experiments using shRNA-target silencing of CD147 to treat transplanted pancreatic cancer tumors in nude mice showed that the tumor size of the experiment group was significantly smaller than that of the control group (39.7mm^3 vs 89.7mm^3), indicating that CD147 as a helper protein could also be used as the target for tumor treatment.

3.12 Urokinase-type plasminogen activator

Urokinase-type plasminogen activator (uPA) is an important component participating in degradation of extracellular matrix and plays an important role in invasion and metastasis of tumor cells. Super-expression of uPA receptor (uPAR or PLAUR) is detected in many malignant tumor cells. Zhou (Zhou *et al,* 2009) used a retroviral vector carrying siRNA expressing target PLAUR to treat transplanted oral squamous cell carcinoma for 30 days, and found that the mean tumor size of the intervention group was 1382mm^3 versus 4181mm^3 in the NS control group. The number of apoptotic cells in the intervention group was significantly greater than that in the control group (32.7 vs 2.7, $P<0.01$). The expression level of u-PAR, MMP2, MMP9, VEGFD, VEGFC and VEGFR-3 in the intervention group decreased significantly, and proliferation-related Ki-67 was inhibited.

3.13 P21-activated kinase 6

Super-expression of P21-activated kinase 6 (encoded by PAK6) is detected in malignant prostate cancer tissues. Wen (Wen *et al,* 2009) compared the inhibitory effect of PAK6-target siRNA (by intratumor injection in transplanted pancreatic cancer tumor-bearing nude mice) and Taxol on tumors, where siRNA was injected into tumors of transplanted pancreatic cancer-bearing nude mice. The result of six-week observation showed that the mean tumor size of the control group, Taxol group, siRNA group and Taxol+siRNA group was 475mm^3, 210mm, 68mm^3 and 47mm^3 respectively. PAK6-siRNA inhibited the growth of pancreatic cancer cells and blocked them at G2-M, demonstrating that both Taxol and PAK6-siRNA could inhibit tumor growth. The inhibitory effect of combined use of Taxol and PAK6-siRNA was significantly better than either of the two along.

3.14 Skp-2

Skp-2 is a substrate recognition subunit of ubiquitin-protein ligase complex SCF/Skp-2, high expression of which in many tumors suggests a poor prognosis. Increased Skp-2 is believed to be associated with cell cycle disturbances in many tumors. HIV retroviral vector-mediated RNAi inhibited Skp-2 high expression cells effectively by increasing the expression of p27 and p21, but it had no significant effect on Skp-2 low expression cells(Sumimoto *et al,* 2005). Intervention of Skp-2 expression in 90% A549 and H1792 lung cancer cell strains could reduce p27 expression, induce apoptosis and inhibit proliferation (Simpson *et al,* 2004). Subcutaneous injection of the adenoviral vector of Skp-2 RNAi could inhibit the growth of subcutaneously transplanted lung cancer effectively, but it was ineffective in tumors without high expression of Skp-2(Faltus *et al,* 2004). Sumimoto *et al* (Sumimoto *et al,* 2006) also found that interference with Skp-2 and BRAFD expression could inhibit the growth and invasion of melanoma cells in vitro.

3.15 Cytokine-induced apoptosis-inhibiting molecule

Cell apoptosis is programmed cell death regulated by multiple genes. Cytokine-induced apoptosis inhibitor 1 (CIAPIN1) is an important regulator of cell apoptosis. It has a significant inhibitory effect on multiple links in cell apoptotic pathways and a positive regulatory effect on Ras and other survival signaling pathways, thus promoting cell survival and proliferation (Shibayama *et al,* 2004). In recent years, many studies on the relationship between CIAPIN1 and tumor formation showed that CIAPIN1 expression was increased significantly in HCC, gastric cancer and other malignant tumor tissues (Ida *et al,* 2007;Zhang *et al,* 2008). Li (Li *et al,* 2008) used adenovirus-mediated RNAi to knock out CIAPIN1 in HCC cells and found that CIAPIN1 gene expression in HCC cells was down-regulated, where cells were prevented from entering phase S and cell proliferation was inhibited. Eight weeks after injection of adenovirus/siRNA between transplanted tumor tissues, the tumor size was only one-sixth of the control group, and pathological sections showed apoptotic cells in tumor tissues.

3.16 Eukaryotic initiation factor 4E

Available studies showed that any form of genetic variation should function at the protein level (Branhart&Simon, 2007). Therefore the relationship between eukaryotic cell (protein translation) initiation factor and tumors has aroused increasing attention, especially eukaryotic initiation factor 4E (eIF4E), which was believed to be more closely related to tumors (Sonenberg,2008). Data have shown that eIF4E can alter the amount of expression of some malignant tumor-related genes and is the determinant of malignant phenotype production. It promotes cells to surpass the limit of normal growth and undergo carcinogenesis through regulating the amount of translation of some specific malignancy-related molecules (Culjkovic *et al,* 2006). Dong (Dong *et al,* 2009) used a vector carrying shRNA of survivin-targeted eIF4E to transfect human breast cancer cells, and found that not only the level of eIF4E mRNA and protein expression was decreased but tumor progression-related protein VEGF, FGF2 and CCND1 were also down-regulated markedly. The breast cancer tumor size of the intervention group using eIF4E-targeted shRNA was significantly smaller than that of the control group ($233.5mm^3$ vs $397.7mm^3$, $P<0.01$). In addition, shRNA could also enhance the chemotherapy effect of cisplatin. The result showed that the tumor size of the cisplatin+shRNA plasmid combination group was significantly smaller than that of the plasmid control group ($134.5mm^3$ vs $208.9mm^3$, $P<0.01$).

3.17 Erythropoietin

Erythropoietin (EPO) was initially found to play a main regulatory role in the proliferation and differentiation of erythroid cells. Studies have shown that t EPO and erythropoietin receptor (EPOR) are expressed in many different nonhematopoietic organs and tissues, playing a role in promoting angiogenesis and protecting tissues. Many recent studies have discovered extensive expressions of EPO and EPOR in many malignant tumors, the autocrine/paracrine pathways of which are associated with tumor microangiogenesis, stimulation of tumor cell proliferation and sensitivity to chemotherapy. In addition, Paragh(Paragh *et al,* 2009) found that EPOR had a function independent of EPO in tumors. It was found that EPOR was highly expressed in ovarian cancer A2780 cell line, but no expression of EPO was observed in normal or hypoxic conditions. Use of exogenous EPO to intervene 2780 cells did not alter their biological effect. Paragh(Paragh *et al,* 2009) transplanted ovarian cancer cells treated with EPOR-target shRNA and those with a negative control vector in mice subcutaneously for 7 weeks, and found that the mean tumor size of the negative control group was 10-fold larger than that of the intervention group. This difference in tumor size is mainly due to decreased cell proliferation in the intervention group. However, the apoptotic rate was similar between the two groups, indicating that EPOR plays a role in regulating cell proliferation, and this effect is independent of EPO.

3.18 miR-221 and miR-222

miR-221 and miR-222 are miRNA in cluster distribution with high homogeneity. Studies have shown that tumor suppressor gene p27 (Fornari *et al,* 2008), p57(Fornari *et al,* 2008), pro-apoptotic factor Bim and Bmf (Gramantieri *et al,* 2009) are all target genes of mir-222/mir-221. studies in recent years have shown that miR-221and miR-222 participate in the development of bladder cancer, HCC, prostate cancer and melanoma. Artificially synthesized small RNA precisely compensatory with miRNA can be used as an miRNA inhibitor to silence endogenous miRNA. Mercatelli *et al* (Mercatelli *et al,* 2008) used anti miR-221 and anti miR-222 inhibitors and a negative control inhibitor to intervene transplanted prostate cancer in nude mice for 33 days, and found that the means tumor size of the intervention group and the control group was 197.2mm^3 and 276.82mm^3 respectively, indicating that regulation of miR221/222 via RNAi could be used as the target of tumor treatment.

3.19 miR-16

miR-16 can target Oncogene BCL-2. By regulating BCL-2 expression with miR-16, tumor cell proliferation can be inhibited. Schaefer *et al* (Schaefer *et al,* 2010) found that miR-16 can down-regulate BCL-2 expression in prostate cancer cells. Takeshita *et al* (Takeshita *et al,* 2010) used chemically synthesized miR-16 to transfect prostate cancer cell line 22Rv1, Du145, PPC-1 and PC-3M-luc, and found that the proliferation of these cells was inhibited. They administered miR-16 and atelocollagen to prevent bone metastasis from transplanted prostate cancer tumor in nude mice via tail vein injection. The results showed that, compared to i.v. injection of atelocollagen as a positive control, miR-16 halted the progression of the transplanted tumors and metastatic lesions, and that miR-16 could regulate the expression of CDK1 and CDK2 that participate in cell cycle and proliferation. Therefore, intravenous administration of miR-16 could be applied as a strategy for the treatment of advanced prostate cancer.

3.20 miR-26a

miR-26a is an miRNA that plays role of tumor suppressor gene. Lu (Lu *et al,* 2011) found that miR-26a could down-regulate the level of EZH2 expression in nasopharyngeal cancer cells, inhibit cell proliferation significantly, and block cell division at G1. Kota *et al* (Kota *et al,* 2009) used miR-26a to intervene HCC, and found that miR-26a could down-regulate the level of cyclin CCND2 and CCNE2 expression, inhibit the proliferation of HCC cells in vitro, and inhibit tumor progression in vivo, indicating that miR-26a could be used as a strategy for tumor treatment by down-regulating cancer genes.

3.21 Specific mutation ectopic gene

1. Ras: Ras is the gene that is most likely to undergo mutation in human malignant tumors. A single-point mutation is enough to transform normal protein into cancer protein. Therefore, there is mild difference in DNA sequence between the wild-type Ras

and mutated-type Ras. Previous methods of gene therapy were unable to inhibit the expression of mutated-type Ras effectively. Researchers have succeeded in using retrovirus-mediated RNA intervention technique to inhibit the expression of mutant K-rasV12 without affecting the other Ras subtypes (Brummelkamp *et al,* 2002). It not only inhibited neoplastic growth and transformation in experiments in vitro but obtained the same result in animal models.

2. bcr-abl: chronic myeloid leukemia (CML) is associated with chromosomal translocation. Protein tyrosine kinase encoded by fusion gene bcr-abl formed durin translocation plays a very important role in the development of CML. Many studies on bcr-abl interference showed that the interference was effective on fusion gene, as shown by down-regulated expression of bcr-abl protein expression, inhibition on cell growth and apoptosis, increased sensitivity to the anti-cancer agent imatinib mesylate, while it had no effect on the wild-type gene in normal tissues. Koldehoff *et al* (Koldehoff *et al,* 2007) first reported the successful use of bcr-abl siRNA in the human body. It enhanced the sensitivity to the anti-cancer agent imatinib mesylate, inhibited the expression of bcr-abl, and promoted cell apoptosis. However, siRNA directing at bcr-abl homologous chromosomes could not enhance the sensitivity to imatinib mesylate and had no effect at all(Lima *et al,* 2004). The reason may be due to different degrees of down-regulation of bcr-Abl protein expression by the siRNA.
3. p53: 50% tumors have mutated-type p53, which attenuates the function of the wild-type. Inhibiting the expression of the mutated-type p53 can help recover the function of the wild-type. A large-scale study on loss of gene function by using RNA Databank(Berns *et al,* 2004) revealed one known and five new p53-dependent hyperplasia suppressor genes (HSG). Inhibition on the expression of these genes can counteract the anti-proliferation function of p53and p19ARF, and thus remove G1 arrest arising from DNA injury.

4. Clinical trials

Local and systemic drug administrations are the main routes of drug delivery in current clinical trials *in vivo*. Local drug administration depends on the type of target tissue, including intravitreal injection for ophthalomological diseases; intranasal administration for respiratory diseases caused by respiratory syncytial virus; intracranial injection for central nervous system diseases; intratumor injection for carcinogenic genes; and intramuscular injection for muscle tissue diseases such as rheumatoid arthritis. Drugs delivered by local administration can directly act on the target organ, thus reducing the probability of degradation of siRNA by nuclease and the dosage of siRNA. Systemic drug administration mainly depends on intraperitoneal and intravenous injection. Systemic drug administration often needs more siRNA to enable the target organ or tissue to obtain an efficacious drug concentration.

A lot of studies have demonstrated that RNAi has substantial potential for tumor treatment. Several RNAi agents have entered clinical trials in the world (Table 1). In May 2008, RNAi agent CALAA-01 for the treatment of solid tumors, developed by Calando Pharmaceuticals, was approved by the Food and Drug Administration (FDA) of the United States for Phase I clinical trial. CALAA-01 is a complex of siRNA and cyclodextrin-transferrin-adamant-PEG nanoparticles, in which human transferrin can specifically interact with transferrin receptor that is overexpressed on the tumor cell membrane, thereby triggering the entry of the nanoparticles into tumor cells through receptor-mediated endocytosis. The target gene is ribonucleotide reductase subunit 2 (RRM2). After systemic administration by i.v. injection, siRNA efficiently inhibited tumor growth by down-regulating RRM2 expression (Davis,2009). It was also found that the levels of RRM2 mRNA and RRM2 protein were both decreased in the tumor. To detect the efficiency of RNAi, researchers found that there existed large amounts of RRM2 mRNA degradation products at tumor site containing the nanoparticles(Davis *et al*, 2010). This is the first clinical trial on RNAi medical treatment of malignant tumors.

SiRNA agent	Company/Institution	Target	Disease	Clinical Trials
CALAA-01	Calando Pharmaceuticals	RRM2	Solid tumor	Phase II
ALN-VSP02	Alnylam	KSP, VEGF	hepatocarcinoma	Phase II
Anti-tat/rev shRNA	Benitec&The city of Hope National Medical Center	Tat & Rev from HIV	AIDS-related lymphoma	Phase I
Atu027	Silence Therapeutics	PKN3	Pulmonary carcinoma	Phase I
BCR-ABL siRNA	University of Duisburg-Essen	Bcr-able	Chronic myeloid leukemia	Single patient
FANG	Gradlis, Inc.	Furin	Advanced cancer	Phase I
SV40/BCR-ABL	Hadassah Medical Organization	Bcr-abl	Chronic myeloid leukemia	Phase I
siRNA immunotherapy	Duke University Hospital	Proteasome	Metastatic melanoma	Phase I

Table 1. A summary of current siRNA agent for clinical trials

Alnylam Pharmaceuticals (Alnylam Pharmaceuticals,2011a,2011b) conducted Phase I clinical trial by intravenous administration of ALN-VSP02 for the treatment of HCC, where the siRNA therapeutic was enveloped in liposome. A total of 58 patients were recruited in the study and administered with the therapeutic at an interval of two weeks. The therapeutic agent targets kinesin spindle protein (KSP) and VEGF that are highly expressed in tumor cells. KSP plays a key role in cell mitosis and is the dynein necessary for bipolar spindle formation and maintenance. Inhibition of KSP would arrest the cell cycle. VEGF promotes angiogenesis, and therefore inhibition of VEGF would suppress tumor growth. Currently, Phase I clinical trial with ALN-VSP02 has been done, and Phase II clinical trial is to be started in 2012.

Recently, Silence Therapeutics has started Phase I clinical trial on Atu027(Aleku et al, 2008; Santel et al, 2010). Atu027 is a lipid-based siRNA complex containing target protein kinase N3 (PKN3), which participates in tumor occurrence and plays an important role in tumor metastasis. Atu027 can inhibit tumor progression by silencing the expression of PKN3. Pre-clinical experiments demonstrated that Atu027 could prevent breast cancer cells from spreading to the lungs in mice. In addition, Atu027 could apparently inhibit the growth of transplanted prostate cancer in nude mice, where the tumor size was about 50% that of the control group; the number of metastatic lymph nodes was only 50% that of the control group; and the lymphatic density in the tumor area was also decreased markedly, though there was no significant difference in vascular density in the tumor area between the study and control groups. In a toxicity experiment, the researchers injected different doses (0.3, 1.0 and 3.0mg/kg) of Atu027 to different groups of cynomolgus monkeys at a 4-day interval, and no significant toxic and adverse effects were observed. Phase I clinical trial with Atu027 mainly showed safety, tolerance and pharmacokinetics.

5. Conclusion

The understanding of signaling pathways participating in the development and progression of malignant tumors and the development of specific target RNAi technique have offered hope to conduct individualized regimens for tumor treatment. However, current RNAi therapeutic approaches are immature and many problems need to be tackled, such as targeting and off-target effects, drug delivery systems, drug administrations and safety. As siRNA with a sequence of only 20-nucleotides could induce a tremendous effect of gene silencing, the sequence of siRNA must be selected accurately in the process of designing siRNA sequences to avoid the off-target effect. Additionally, a mRNA target sequence may hide in the mRNA secondary structure or folding region, making it impossible for siRNA to silence the mRNA. It is therefore necessary to conduct experiments repeatedly to select a suitable target site. Before RNAi medical treatment can be used clinically, it is necessary to seek an efficient delivery system and rational drug administration. In addition, as nucleotide sequences used for RNAi have no cell targetability and are likely to degrade, it is necessary to design and synthesize a drug delivery system that is non- or low toxic to normal cells and can deliver siRNA to tumor cells efficiently.

6. Acknowledgment

This work was financially supported by the Key Scientific Project of Shanghai Putuo District(PTKW10-B01), the Young Program of Shanghai health bureau(2009Y127), Young backbone promotion plan of Shanghai society of Nuclear Medicine (2009-NM-07).

7. References

Aleku, M.; Schulz, P.; Keil, O.; Santel, A.; Schaeper, U.; Dieckhoff, B.; Janke, O.; Endruschat, J.; Durieux, B.; Roder, N.; Loffler, K.; Lange, C.; Fechtner, M.; Mopert, K.; Fisch, G.; Dames, S.; Arnold, W.; Jochims, K.; Giese, K.; Wiedenmann, B.; Scholz, A.; Kaufmann, J.(2008). Atu027, a liposomal small interfering RNA formulation targeting protein kinase N3, inhibits cancer progression, *Cancer Research.*, Vol.68, No.23, (December 2008), pp.9788-9798, ISSN0008-5472.

Alnylam Pharmaceuticals. (August 2011).Dose Escalation trial to evaluate the safety, tolerability, pharmacokinetics and pharmacodynamics of intravenous ALN-VSP02 in patients with advanced solid tumors with liver involvement, 23.08.2011, Available from http://clinicaltrials.gov/show/NCT00882180.

Alnylam Pharmaceuticals. (June 2011).Alnylam Presents Phase I Data For ALN-VSP, An RNAi Therapeutic For The Treatment Of Liver Cancers, At ASCO Meeting, 06.06.2011, Available from http://www.medicalnewstoday.com/releases/227569.php

Berns, K.; Hijmans, E.; Mullenders, J.; Brummelkamp, T.; Velds, A.; Heimerikx, M.; Kerkhoven, R.; Madiredjo, M.; Nijkamp, W.; Weigelt, B.; Agami, R.; Ge, W.; Cavet, G.; Linsley, P.; Beijersbergen, R.; Bernards, R.(2004). A large-scale RNAi screen in human cells identifies new components of the p53 pathway, *Nature*, Vol.428, No.6981, (March 2004), pp.431-437, ISSN 0028-0836

Berns, K.; Hijmans, E.; Mullenders, J.; Brummelkamp, T.; Velds, A.; Heimerikx, M.; Kerkhoven, R.; Madiredjo, M.; Nijkamp, W.; Weiqelt, B.; Aqami, R.; Ge, W.; Cavet, G.; Linsley, P.; Beijersberqen, R.; Bernards, R.(2004). A large-scale RNAi screen in human cells identifies new components of the p53 pathway, *Nature*, Vol.428, No.6981, (March 2004), pp.431-437, ISSN 0028-0836.

Bernstein, E.; Caudy, A.; Hammond, S.; Hannon, G.(2001). Role for a bidentate ribonuclease in the initiation step of RNA interference, *Nature*, Vol.409, No. 6818, (January 2001), pp.363-366, ISSN 0028-0836

Blaszczyk, J.; Tropea, J.; Bubunenko, M.; Routzahn, K.; Waugh, D.; Court, D.; Ji, X. (2001). Crystallographic and modeling studies of RNase III suggest a mechanism for double-stranded RNA cleavage. *Structure*, Vol.79, No. 12, (December 2001), pp.1225-1236, ISSN 0969-2126

Branhart, B.; Simon, M.(2007). Taking aim at translation for tumor therapy, *The Journal of clinical investigation* , Vol.117, No.9, (September 2007), pp.2385-2388, ISSN 0021-9738.

Brummelkamp, T.; Bernards , R.; Agami , R.(2002). Stable suppression of tumorigenicity by virus-mediated RNA interference, *Cancer Cell*, Vol.2, No.3, (September 2002), pp.243-247, ISSN 0007-9235.

Brummelkamp, T.; Nijman, S.; Dirac, A.; Bernards, R. (2003).Loss of the cylindromatosis tumour suppressor inhibits apoptosis by activating NF-kappaB, *Nature,* Vol.424, No.6950, (August 2003), pp.797-801, ISSN 0028-0836

Chen, T.; Deng, C.(2008). Inhibitory effect of siRNA targeting survivin in gastric cancer MGC-803 cells, *International immunopharmacology*, Vol.8, No.7, (July 2008), pp.1006-1011, ISSN 1567-5769

Collins, R.; Cheng, X. (2005).Structural domains in RNAi, *FEBS Lett*, Vol.579, No. 26, (October 2005), pp.5841-5849, ISSN 0014-5793

Coppola, D.; Szabo, M.; Boulware, D.; Muraca, P.; Alsarraj, M.; Chambers, A.; Yeatman, T.(2004). Correlation of osteopontin protein expression and pathological stage across a wide variety of tumor histologies, *Clinical cancer research*, Vol.10, No.1(Pt1), (January 2004), pp.184-190, ISSN 1078-0432

Cordin, O.; Banroques, J.; Tanner, N.; Linder, P.(2006) The DEAD-box protein family of RNA helicases. *Gene*, Vol.367, (February 2006), pp.17-37, ISSN 0378-1119.

Cui, R.; Takahashi, F.; Ohashi, R.; Gu, T.; Yoshioka, M.; Nishio, K.; Ohe, Y.; Tominaga, S.; Takagi, Y.; Sasaki, S.; Fukuchi, Y.; Takahashi, K.(2007). Abrogation of the interaction between osteopontin and alphavbeta3 integrin reduces tumor growth of human lung cancer cells in mice, *Lung cancer*, Vol.57, No.3, (September 2007), pp.302-310, ISSN 0169-5002

Culjkovic, B.; Topisirovic, I.; Skrabanek, L.; Ruiz-Gutierrez, M.; Borden, K.(2006). eIF4E is a central node of an RNA regulon that governs cellular proliferation, *The Journal of cell biology*, Vol.175, No.3, (November 2006), pp.415-426, ISSN 0021-9525.

Davis, M.(2009). The first targeted delivery of siRNA in humans via a self-assembling, cyclodextrin polymer-based nanoparticle: from concept to clinic, *Molecular Pharmaceutics*, Vol.6, No.3, (May-June 2009), pp.659-668, ISSN 1543-8384.

Davis, M.; Zuckerman, J.; Choi, C.; Seligson, D.; Tolcher, A.; Alabi, C.; Yen, Y.; Heidel, J.; Ribas, A. (2010).Evidence of RNAi in humans from systemically administered siRNA via targeted nanoparticles, *Nature*, Vol.464, No. 7291, (April 2010), pp. 1067-1070, ISSN 0028-0836

Davis, M.; Zuckerman, J.; Choi, C.; Seligson, D.; Tolcher, A.; Alabi, C.; Yen, Y.; Heidel, J.; Ribas, A.(2010). Evidence of RNAi in humans from systemically administered siRNA via targeted nanoparticles, *Nature*, Vol.464, No.7291, (April 2010), pp.1067-1070, ISSN 0028-0836.

de Souza Nascimento, P.; Alves, G.; Fiedler, W.(2006). Telomerase inhibition by an siRNA directed against hTERT leads to telomere attrition in HT29 cells, *Oncology reports*, Vol.16, No.2, (August 2006), pp.423-428, ISSN 1021-335X

Devi, G.(2006). siRNA-based approaches in cancer therapy, *Cancer gene therapy*, Vol.13, No.9, (September 2006), pp.819-829, ISSN 0929-1903

Dong, K.; Wang, R.; Wang, X.; Lin, F.; Shen, J.; Gao, P.; Zhang, H.(2009).Tumor-specific RNAi targeting eIF4E suppresses tumor growth, induces apoptosis and enhances cisplatin cytotoxicity in human breast carcinoma cells, *Breast cancer research and treatment*, Vol.113, No.3, (February 2009), pp.443-456, ISSN 0167-6806.

Du, W.; Wang, L.; Cao, S.; Chen, B.; Zhang, Y.; Bai, F.; Liu, J.; Fan, D.(2009). EpCAM is overexpressed in gastric cancer and its downregulation suppresses proliferation of gastric cancer, *Journal of cancer research and clinical oncology*, Vol.135, No.9, (September 2009), pp.1277-1285, ISSN 0171-5216

Faltus, T.; Yuan, J.; Zimmer, B.; Krämer, A.; Loibl, S.; Kaufmann, M.; Strebhardt, K.(2004). Silencing of the HER2/neu gene by siRNA inhibits proliferation and induces apoptosis in HER2/neu-overexpressing breast cancer cells, *Neoplasia*, Vol.6, No.6, (November-December 2004), pp.786-795, ISSN 1522-8002

Fire, A.; Xu, S.; Montgomery, M.; Kostas, S.; Dreiver, S.; Mello, C.(1998). Potent and specific genetic interference by double-stranded RNA in Caenorhabditis elegans, *Nature*, Vol.391, No.6669, (February 1998), pp. 806-811, ISSN 0028-0836.

Fornari, F.; Gramantieri, L.; Ferracin, M.; Veronese, A.; Sabbioni, S.; Calin, G.; Grazi, G.; Giovannini, C.; Croce, C.; Bolondi, L.; Neqrini, M.(2008). MiR-221 con-trols CDKN1C/p57 and CDKN1B/p27 expression in human hepatocellular carcinoma, *Oncogene*, Vol.27, No.43, (September 2008), pp.5651-5661, ISSN 0950-9232.

Gandellini, P.; Folini, M.; Bandiera, R.; De Cesare, M.; Binda, M.; Veronese, S.; Daidone, M.; Zunino, F.; Zaffaroni, N.(2007). Down-regulation of human telomerase reverse transcriptase through specific activation of RNAi pathway quickly results in cancer cell growth impairment, *Biochemical pharmacology*, Vol.73, No.11, (June 2007), pp.1703-1714, ISSN 0006-2952

Gong, M.; Lu, Z.; Fang, G.; Bi, J.; Xue, X.(2008). A small interfering RNA targeting osteopontin as gastric cancer therapeutics, Cancer letters, Vol.272, No.1, (December 2008), pp.148-159, ISSN 0304-3835

Gramantieri, L.; Fornari, F.; Ferracin, M.; Veronese, A.; Sabbioni, S.; Calin, G.; Grazi, G.; Croce, C.; Bolondi, L.; Neqrini, M. (2009).MicroRNA-221 targets Bmf in hepatocellular carcinoma and correlates with tumor multifocality, *Clinical cancer research*, Vol.15, No.16, (August 2009), pp.5073-5081, ISSN 1078-0432.

Gregory, R.; Chendrimada, T.; Cooch, N.; Shiekhattar, R.(2005). Human RISC couples microRNA biogenesis and posttranscriptional gene silencing, *Cell*, Vol.123, No.4, (November 2005), pp.632-640, ISSN 0092-8674

Guo, S.; Kemphues, K. (1995).par-1, a gene required for establishing polarity in C. elegans embryos, encodes a putative Ser/Thr kinase that is asymmetrically distributed, *Cell*, Vol.81, No. 4, (May 1995), pp.611-620, ISSN 0092-8674

Hallegger, M.; Taschner, A.; Jantsch, M. (2006).RNA aptamers binding the double-stranded RNA-binding domain.*RNA*, Vol.12, No. 11, (November 2006), pp.1993-2004, ISSN 1355-8382

Hammond, S.; Bernstein, E.; Beach, D.; Hannon, G.(2000). An RNA-directed nuclease mediates post-transcriptional gene silencing in Drosophila cells, *Nature*, Vol.404, No.6775, (March 2000), pp.293-296, ISSN 0028-0836

Hammond, SM. (2005).Dicing and slicing: the core machinery of the RNA interference pathway, *FEBS Letters*, Vol.579, No.26, (October 2005), pp.5822-5829, ISSN 0014-5793

Hannon, G. (2002).RNA interference, *Nature*, Vol.418, No.6894, (July 2002), pp. 244-251, ISSN 0028-0836.

He, X.; Liu, T.; Xiao, Y.; Feng, Y.; Cheng, D.; Ting, G.; Zhang, L.; Zhang, Y.; Chen, Y.; Ting, G.; Zhang, L.(2009). Vascular endothelial growth factor-C siRNA delivered via calcium carbonate nanoparticle effectively inhibits lymphangiogenesis and growth of colorectal cancer in vivo, *Cancer biotherapy & radiopharmaceuticals*, Vol.24, No.2, (April 2009), pp.249-259, ISSN 1084-9785

Huang, C.; Jiang, T.; Zhu, L.; Liu, J.; Cao, J.; Huang, K.; Qiu, Z.(2011). STAT3-targeting RNA interference inhibits pancreatic cancer angiogenesis in vitro and in vivo, *International journal of oncology*, Vol.38, No.6, (June 2011), pp.1637-1644, ISSN 1019-6439

Ida, H.; Yoshida, H.; Nakamura, K.; Yamaguchi, M.(2007). Identification of the Drosophila eIF4A gene as a target of the DREF transcription factor, *Experimental cell research*, Vol.313, No.20, (December 2007), pp.4208-4220, ISSN 0014-4827

Jemal, A.; Bray, F.; Center, M.; Ferlay, J; Ward, E.; Forman, D. (2011).Global cancer statistics, *Cancer Journal for Clinicians*, Vol.6, No.4, (Mar-Apr 2011), pp. 69-90, ISSN 0007-9235.

Kanwar, J.; Kamalapuram, S.; Kanwar, R.(2010). Targeting survivin in cancer: patent review, *Expert opinion on therapeutic patents,* Vol.20, No.12, (December 2010), pp.1723-1737, ISSN 1354-3776

Koldehoff , M.; Steckel , N.; Beelen , D.; Elmaaqacli, A.; (2007). Therapeutic application of small interfering RNA directed against bcr-abl transcripts to a patient with imatinib-resistant chronic myeloid leukaemia, *Clinical and Experimental Medicine,* Vol.7, No.2, (June 2007), pp.47-55, ISSN 1591-8890.

Koldehoff, M.; Elmaagacli, A. (2009).Therapeutic targeting of gene expression by siRNAs directed against BCR-ABL transcripts in a patient with imatinib-resistant chronic myeloid leukemia, *Methods in molecular biology,* Vol.487, (2009), pp. 451-466, ISSN 1064-3745

Kota, J.; Chivukula, R.; O'Donnell, K.; Wentzel, E.; Montqomery, C.; Hwanq, H.; Chang, T.; Vivekanandan, P.; Torbenson, M.; Clark, K.; Mendell, J.; Mendell, J.(2009).Therapeutic microRNA delivery suppresses tumorigenesis in a murine liver cancer model, *Cell,* Vol.137, No.6, (June 2009), pp.1005-1017, ISSN 0092-8674.

Krek, A.; Grün, D.; Poy, M.; Wolf, R.; Rosenberg, L.; Epstein, E.; MacMenamin, P.; da Piedade, I.; Gunsalus, K.; Stoffel, M.; Rajewsky, N. (2005).Combinatorial microRNA target predictions, *Nature genetics,* Vol.37, No.5, (May 2005), pp.495-500, ISSN 1061-4036

Li, J.; Piao, Y.; Jiang, Z.; Chen, L.; , Sun, H. (2009) . Silencing of signal transducer and activator of transcription 3 expression by RNA interference suppresses growth of human hepatocellular carcinoma in tumor-bearing nude mice, *World journal of gastroenterology,* Vol.15, No.21, (June 2009), pp.2602-2608, ISSN 1007-9327

Li, X.; Pan, Y.; Fan, R.; Jin, H.; Han, S.; Liu, J.; Wu, K.; Fan, D.(2008). Adenovirus-delivered CIAPIN1 small interfering RNA inhibits HCC growth in vitro and in vivo, *Carcinogenesis,* Vol.29, No.8, (August 2008), pp.1587-1593, ISSN 0143-3334

Lima, R.; Martins, L.; Guimaraes, J.; Sambade, C.; Vasconcelos, M. (2004). Specific downregulation of bcl-2 and xIAP by RNAi enhances the effects of chemotherapeutic agents in MCF-7 human breast cancer cells, *Cancer Gene Therapy,* Vol.11, No.5, (March 2004), pp.309-316, ISSN0929-1903.

López-Fraga, M.; Martínez, T.; Jiménez, A. (2009).RNA interference technologies and therapeutics: from basic research to products. BioDrugs, Vol.23, No.5, (October 2009), pp. 305-332, ISSN 1173-8804.

Lu, J.; He, M.; Wang , L.; Chen, Y.; Liu, X.; Dong, Q.; Chen, Y.; Peng, Y.; Yao, K.; Kunq, H.; LI, X.; (2011). MiR-26a Inhibits Cell Growth and Tumorigenesis of Nasopharyngeal Carcinoma through Repression of EZH2, *Cancer research.,* Vol.71, No.1, (January 2011), pp.225-233, ISSN 0008-5472.

Ma, J.; Ye, K.; Patel, D. (2004).Structural basis for overhang-specific small interfering RNA recognition by the PAZ domain. *Nature,* Vol.429, No. 6989, (May 2004), pp.318-322, ISSN 0028-0836

Ma, J.; Yuan, Y.; Meister, G.; Pei, Y.; Tuschl, T.; Patel, D.(2005). Structural basis for 5'-end-specific recognition of guide RNA by the A. fulgidus Piwi protein, *Nature,* Vol.434, No.7033, (March 2005), pp.666-670, ISSN 0028-0836

Macrae, I.; Li, F.; Zhou, K.; Cande, W.; Doudna, J.(2006). Structure of Dicer and echanistic implications for RNAi, *Cold Spring Harbor symposia on quantitative biology*, Vol.71, (2006), pp.73-80, ISSN 0091-7451

Maples, P.; Kumar, P.; Yu, Y.; Wang, Z.(2009).FANG vaccine: autologous tumor vaccine genetically modified to express GMCSF and block production of furin, BioProcessing Journal, Vol.8, (2009), pp. 4-14, ISSN 1538-8786

Matranga, C.; Tomari, Y.; Shin, C.; Bartel, D.; Zamore, P. (2005).Passenger-strand cleavage facilitates assembly of siRNA into Ago2-containing RNAi enzyme complexes, *Cell*, Vol.123, No.4, (November 2005), pp.607-620, ISSN 0092-8674

Meister G, Landthaler M, Peters L, Chen PY, Urlaub H, Lührmann R, Tuschl T.(2005). Identification of novel argonaute -associated proteins, *Current biology*, Vol.15, No.23, (December 2005), pp.2149-2155, ISSN 0960-9822

Meister, G.; Landthaler, M.; Patkaniowska, A.; Dorsett, Y.; Teng, G.; Tuschl, T.(2004). Human Argonaute2 mediates RNA cleavage targeted by miRNAs and siRNAs, *Molecular cell*, Vol.15, No.2, (July 2004), pp.185-197, ISSN 1097-2765

Meister, G.; Tuschl, T. (2004).Mechanisms of gene silencing by double-stranded RNA, *Nature*, Vol.431, No.7006, (September 2004), pp.343-339, ISSN 0028-0836

Mercatelli, N.; Coppola, V.; Bonci, D.; Miele, F.; Costantini, A.; Guadagnoli, M.; Bonanno, E.; Muto, G.; Frajese, G.; De, R.; Spagnoli, L.; Farace, M.; Ciafre, S.(2008). The inhibition of the highly expressed miR-221 and miR-222 impairs the growth of prostate carcinoma xenografts in mice, *PLoS One*, Vol.3, No.12, (December 2008), pp.4029, ISSN 1932-6203.

Paccione, R.; Miyazaki, H.; Patel, V.; Waseem, A.; Gutkind, J.; Zehner, Z.; Yeudall, W.(2008). Keratin down-regulation in vimentin-positive cancer cells is reversible by vimentin RNA interference, which inhibits growth and motility, *Molecular cancer therapeutics*, Vol.7, No.9, (September 2008), pp.2894-2903, ISSN 1535-7163

Paddison, P.; Silva, J.; Conklin, D.; Schlabach, M.; Li, M.; Aruleba, S.; Balija, V.; O'Shaughnessy, A.; Gnoj, L.; Scobie, K.; Chang, K.; Westbrook, T.; Cleary, M.; Sachidanandam, R.; McCombie, W.; Elledge, S.; Hannon, G.(2004). A resource for large-scale RNA-interference-based screens in mammals, *Nature*, Vol.428, No.6981, (March 2004), pp.427-431, ISSN 0028-0836

Paragh , G.; Kumar, S.; Rakosy, Z.; Choi, S.; Xu, X.; Acs, G.(2009). RNA interference-mediated inhibition of erythropoietin receptor expression suppresses tumor growth and invasiveness in A2780 human ovarian carcinoma cells, *The American journal of pathology*, Vol.174, No.4, (April 2009), pp.1504-1514, ISSN 0002-9440.

Pillai, R.; Bhattacharyya, S.; Filipowicz, W.(2007). Repression of protein synthesis by miRNAs: how many mechanisms? *Trends in cell biology*, Vol.17, No.3, (March 2007), pp.118-126, ISSN 0962-8924

Preall, JB.; Sontheimer, EJ.(2005). RNAi: RISC gets loaded, *Cell*, Vol.123, No.4, (November 2005), pp.543-545, ISSN 0092-8674

Ramaiah, S.; Rittling, S.(2008). Pathophysiological role of osteopontin in hepatic inflammation, toxicity, and cancer, *Toxicological sciences*, Vol.103, No.1, (May 2007), pp.4-13, ISSN 1096-6080

Rao, D.; Maples, P.; Senzer, N.; Kumar, P.; Wang, Z.; Pappen, B.; Yu, Y.; Haddock, C.; Jay, C.; Phadke, A.; Chen, S.; Kuhn, J.; Dylewski, D.; Scott, S.; Monsma, D.; Webb, C.; Tong, A.; Shanahan, D.; Nemunaitis, J.(2010). Enhanced target gene knockdown by a bifunctional shRNA: a novel approach of RNA interference, *Cancer gene therapy*, Vol.17, No.11, (November 2010), pp.780-791, ISSN 0929-1903

Rao, D.; Vorhies, J.; Senzer, N.; Nemunaitis, J.(2009). siRNA vs. shRNA: similarities and differences, *Advanced drug delivery reviews*, Vol.61, No.9, (July 2009), pp.746-759, ISSN 0169-409X

Robb, G.; Rana, T. (2007).RNA helicase A interacts with RISC in human cells and functions in RISC loading, *Molecular cell*, Vol.26, No.4, (May 2007), pp.523-537, ISSN 1097-2765

Ryter, J.; Schultz, S. (1998).Molecular basis of double-stranded RNA-protein interactions: structure of a dsRNA-binding domain complexed with dsRNA, *The EMBO journal*, Vol.17, No.24, (December 1998), pp.7505-7513, ISSN 0261-4189

Santel, A.; Aleku, M.; Röder, N.; Mopert, K.; Durieux, B.; Janke, O.; Keil, O.; Endruschat, J.; Dames, S.; Lange, C.; Eisermann, M.; Loffler, K.; Fechtner, M.; Fisch, G.; Vank, C.; Schaeper, U.; Giese, K.; Kaufmann, J. (2010). Atu027 prevents pulmonary metastasis in experimental and spontaneous mouse metastasis models, *Clinical Cancer Research*, Vol.16, No.22, (November 2010), pp.5469-5480, ISSN 1078-0432.

Schaefer, A.; Jung, M.; Mollenkopf, H.; Waqner, I.; Stephan, C.; Jentzmik, F.; Miller, K.; Lein, M.; Kristiansen, G.; Junq, K.(2010), Diagnostic and prognostic implications of microRNA profiling in prostate carcinoma, *International journal of cancer*, Vol.126, No.5, (March 2010), pp.1167-1176, ISSN 0020-7136.

Schneiderhan, W.; Scheler, M.; Holzmann, K.; Marx, M.; Gschwend, J.; Bucholz, M.; Gress, T.; Seufferlein, T.; Adler, G.; Oswald, F.(2009). CD147 silencing inhibits lactate transport and reduces malignant potential of pancreatic cancer cells in in vivo and in vitro models, *Gut*, Vol.58, No.10, (October 2009), pp.1391-1398, ISSN 0017-5749

Schwarz, D.; Tomari, Y.; Zamore, P. (2004).The RNA-induced silencing complex is a Mg2+-dependent endonuclease, *Current biology*, Vol.14, No.9, (May 2004), pp.787-791, ISSN 0960-9822

Seth, S.; Matsui, Y.; Fosnaugh, K.; Liu, Y.; Vaish, N.; Adami, R.; Harvie, P.; Johns, R.; Severson, G.; Brown, T.; Takagi, A.; Bell, S.; Chen, Y.; Chen, F.; Zhu, T.; Fam, R.; Maciagiewicz, I.; Kwang, E.; McCutcheon, M.; Farber, K.; Charmley, P.; Houston, M.; So, A.; Templin, M.; Polisky, B. (2011).RNAi-based therapeutics targeting survivin and PLK1 for treatment of bladder cancer, *Molecular therapy* , Vol.19, No.5, (May 2011), pp.928-935, ISSN 1525-0016

Sethupathy, P.; Corda, B.; Hatzigeorgiou, AG. (2006).TarBase: A comprehensive database of experimentally supported animal microRNA targets, *RNA*, Vol.12, No.2, (February 2006), pp.192-197, ISSN 1355-8382

Shibayama, H.; Takai, E.; Matsumura, I.; Kouno, M.; Morii, E.; Kitamura, Y.; Takeda, J.; Kanakura, Y.(2004). Identification of a cytokine-induced antiapoptotic molecule anamorsin essential for definitive hematopoiesis, *The Journal of experimental medicine*, Vol.199, No.4, (Febuary 2004), pp.581-592, ISSN 0022-1007

Silva, J.; Li, M.; Chang, K.; Ge, W.; Golding, M.; Rickles, R.; Siolas, D.; Hu, G.; Paddison, P.; Schlabach, M.; Sheth, N.; Bradshaw, J.; Burchard, J.; Kulkarni, A.; Cavet, G.; Sachidanandam, R.; McCombie, W.; Cleary, M.; Elledge, S.; Hannon, G.(2005). Second-generation shRNA libraries covering the mouse and human genomes, *Nature genetics*, Vol.37, No.11, (November 2005), pp.1281-1288, ISSN 1061-4036

Simpson, K.; Dugan, A.; Mercurio, A.(2004). Functional analysis of the contribution of RhoA and RhoC GTPases to invasive breast carcinoma, *Cancer research*, Vol.64, No.23, (December 2004), pp.8694-8701, ISSN 0008-5472

Sonenberg, N.(2008). eIF4E, the mRNA cap-binding protein: from basic discovery to translational research, *Biochemistry and cell biology*, Vol.86, No.2, (April 2008), pp.178-183, ISSN 0829-8211

Song, J.; Smith, S.; Hannon, G.; Joshua-Tor, L. (2004).Crystal structure of Argonaute and its implications for RISC slicer activity, *Science*, Vol.305, No.5689, (September 2004), pp.1434-1437, ISSN 0036-8075

Sumimoto, H.; Hirata, K.; Yamagata, S.; Miyoshi, H.; Miyagishi, M.; Taira, K.; , Kawakami, Y.(2006). Effective inhibition of cell growth and invasion of melanoma by combined suppression of BRAF (V599E) and Skp2 with lentiviral RNAi, *International journal of cancer*, Vol.118, No.2, (January 2006), pp.472-476, ISSN 0020-7136

Sumimoto, H.; Yamagata, S.; Shimizu, A.; Miyoshi, H.; Mizuguchi, H.; Hayakawa, T.; Miyagishi, M.; Taira, K.; Kawakami, Y.(2005). Gene therapy for human small-cell lung carcinoma by inactivation of Skp-2 with virally mediated RNA interference, *Gene therapy*, Vol.12, No.1, (January 2005), pp.95-100, ISSN 0969-7128

Takeshita, F.; Patrawala, L.; Osaki, M.; Takahashi, R.; Yamamoto, Y.; Kosaka, N.; Kawamata, M.; Kelnar, K.; Bader, A.; Browm, D.; Ochiya, T. (2010). Systemic delivery of synthetic microRNA-16 inhibits the growth of metastatic prostate tumors via downregulation of multiple cell-cycle genes, *Molecular Therapy*, Vol.18, No.1, (January 2010), pp.181-187, ISSN 1525-0016.

Tang, G. (2005).siRNA and miRNA: an insight into RISCs, *Trends in biochemical sciences*, Vol.30, No.2, (February 2005), pp.106-114, ISSN 0968-0004

Tang, H.; Wang, J.; Bai, F.; Hong, L.; Liang, J.; Gao, J.; Zhai, H.; Lan, M.; Zhang, F.; Wu, K.; Fan, D.(2007). Inhibition of osteopontin would suppress angiogenesis in gastric cancer, *Biochemistry and cell biology*, Vol.85, No.1, (Febuary 2007), pp.103-110, ISSN 0829-8211

Tomari, Y.; Matranga, C.; Haley, B.; Martinez, N.; Zamore, P.(2004). A protein sensor for siRNA asymmetry, *Science*, Vol.306, No.5700, (November 2004), pp.1377-1380, ISSN 0036-8075

Wai PY, Kuo PC.(2008). Osteopontin: regulation in tumor metastasis, *Cancer metastasis reviews*, Vol.27, No.1, (March 2008), pp.103-108, ISSN 0167-7659

Wang, J.; Ji, A.; Wen, J.; Ren H.(2010). Effect of β-catenin gene silencing by shRNA on biologic characteristics of human esophageal carcinoma cells, *Chinese Journal of Pathology*, Vol.39, No.12, (December 2010), pp.835-841, ISSN 0529-5807

Wang, J.; Zheng, C.; Wang, Y.; Wen, J.; Ren, H.; Liu, Y.; Jiang, H.(2008).Gene silencing of beta-catenin by RNAi inhibits cell proliferation in human esophageal cancer cells in vitro and in nude mice, *Diseases of the esophagus,* Vol.22, No.2, (April 2009), pp.151-162, ISSN 1120-8694

Wang, Q.; Lv, Y.; Fei, L.(2010). New Thoughts on Cancer Gene Therapy, *Chinese Journal of Clinical Oncology,* Vol.37, No.15, (August 2010), pp. 893-896, ISSN 1000-8179.

Wen, X.; Li, X.; Liao, B.; Liu, Y.; Wu, J.; Yuan, X.; Ouyang, B.; Sun, Q.; Gao, X.(2009). Knockdown of p21-activated kinase 6 inhibits prostate cancer growth and enhances chemosensitivity to docetaxel, *Urology,* Vol.73, No.6, (June 2009), pp.1407-1411, ISSN 0090-4295

Wise, G.; King, G.(2008). Mechanisms of tooth eruption and orthodontic tooth movement, *Journal of dental research,* Vol.87, No.5, (May 2008), pp.414-434, ISSN 0022-0345

Wu, H.; Henras, A.; Chanfreau, G.; Feigon, J. (2004).Structural basis for recognition of the AGNN tetraloop RNA fold by the double-stranded RNA-binding domain of Rnt1p RNase III. *Proceedings of the National Academy of Sciences of the United States of America,* Vol.101, No.22, (June 2004), pp.8307-8312, ISSN 0027-8424

Xv, N.; Dong, Z.(2003). Overview of the Cancer Epidemiological Status in China and Strategy on Cancer Control, *China Journal of Cancer Prevention and treatment,* Vol.10, No.1, (January 2003), pp. 71-74, ISSN 1673-5269.

Yuan, Y.; Pei, Y.; Ma, J.; Kuryavyi, V.; Zhadina, M.; Meister, G.; Chen, H.; Dauter, Z.; , Tuschl, T.; Patel, D. (2005).Crystal structure of A. aeolicus argonaute, a site-specific DNA-guided endoribonuclease, provides insights into RISC-mediated mRNA cleavage, *Molecular cell,* Vol.19, No.3, (August 2005), pp.405-419, ISSN 1097-2765

Zeng, P.; Wagoner, H.; Pescovitz, O.; Steinmetz, R.(2005). RNA interference (RNAi) for extracellular signal-regulated kinase 1 (ERK1) alone is sufficient to suppress cell viability in ovarian cancer cells, *Cancer biology & therapy,* Vol.4, No.9, (September 2005), pp.961-967, ISSN 1538-4047

Zhang, W.; Erkan, M.; Abiatari, I.; Giese, N.; Felix, K.; Kayed, H.; Büchler, M.; Friess, H.; Kleeff, J. (2007). Expression of extracellular matrix metalloproteinase inducer (EMMPRIN/CD147) in pancreatic neoplasm and pancreatic stellate cells, *Cancer biology & therapy,* Vol.6, No.2, (February 2007), pp.218-227, ISSN 1538-4047

Zhang, X.; Ge, Y.; Tian, R.(2009). The knockdown of c-myc expression by RNAi inhibits cell proliferation in human colon cancer HT-29 cells in vitro and in vivo, *Cellular & molecular biology letters,* Vol.14, No.2, (January 2009), pp.305-318, ISSN 1425-8153

Zhang, X.; Ladd, A.; Dragoescu, E.; Budd, W.; Ware, J.; Zehner, Z.(2009). MicroRNA-17-3p is a prostate tumor suppressor in vitro and in vivo, and is decreased in high grade prostate tumors analyzed by laser capture microdissection, *Clinical & experimental metastasis,* Vol.26, No.8, (September 2009), pp.965-979, ISSN 0262-0898

Zhang, Y.; Lyver, E.; Nakamaru-Ogiso, E.; Yoon, H.; Amutha, B.; , Lee, D.; Bi, E.; Ohnishi, T.; Daldal, F.; Pain, D.; Dancis, A.(2008). Dre2, a conserved eukaryotic Fe/S cluster protein, functions in cytosolic Fe/S protein biogenesis, *Molecular and cellular biology,* Vol.28, No.18, (September 2008), pp.5569-5582, ISSN 0270-7306

Zhang, B.; Zhang, Z.; Wang, C.(2010). Inhibitory effects on proliferation and invasion of lung cancer cells by RNAi-mediated knockdown of osteopontin, *Chinese Journal of experimental surgery,* Vol.27, No.2, (Febuary 2010), pp.230-231, ISSN 1001-9030

Zhang, S.; Chen, W.; Lei Z.; Zou, X.; Zhao, P. (2008).A Report of Cancer Incidence from 37 Cancer Registries in China, 2004, *Bulletin of Chinese Cancer,* Vol.17, No.11, (November 2011), pp. 909-912, ISSN 1004-0242.

Zhou, H.; Tang, Y.; Liang, X.; Yang, X.; Yang, J.; Zhu, G.; Zheng, M.; Zhang, C.(2009). RNAi targeting urokinase-type plasminogen activator receptor inhibits metastasis and progression of oral squamous cell carcinoma in vivo, *International journal of cancer,* Vol.125, No.2, (July 2009), pp.453-462, ISSN 0020-7136

4

Bioreducible Cationic Polymers for Gene Transfection

Chao Lin and Bo Lou
Tongji University School of Medicine
The Institute for Advanced Materials and Nanobiomedicine
Tongji University, Shanghai
P.R. China

1. Introduction

Gene therapy holds substantial promise for the treatment of a broad class of overwhelming human diseases such as cancer and AIDS (Verma & Somia, 1997). An essential procedure in gene therapy program involves the delivery of encoded plasmid genes into the patient's somatic cells so as to express therapeutic proteins. An ideal strategy for successful gene delivery depends on safe and efficient gene delivery vectors (El-Aneed, 2004). Generally, gene delivery vectors are classified into two categories: viral vectors and non-viral vectors. Viral vectors are derived from natural viruses such as adenovirus and retrovirus with eliminated pathogenicity. Because of their unique capability in cell infection, viral vectors are most popular for gene delivery *in vitro* and *in vivo*. Unfortunately, clinical practice of viral vectors is seriously hampered by a few inherent issues including random insertion into the host genomes, immunogenicity, gene-carrying capacity limitation, and small-scale production (C.E. Thomas et al., 2003). In the past decades, these safety concerns on viral vectors have led to accelerated advancement in non-viral vectors (S. Li & Huang, 2000). Non-viral vectors such as lipids and polymers take more advantages over conventional viral vectors, including low immunogenicity after repeated administration, easy manufacture, large-scale production and low cost. However, current non-viral systems typically fail to give rise to as efficient gene transfection as powerful viral vectors (Pack et al., 2005). Thus, the availability of highly potent non-viral gene delivery vectors still remains a big challenge.

Among different non-viral vectors, cationic polymers have received much attention because they can be prepared by different polymerization methods and easily modified to introduce different bio-functional groups (Luo & Saltzman, 2000). In the past two decades, a few traditional cationic polymers such as chitosan, polyethylenimine (pEI), poly(L-lysine) (pLL), polyamidoamine (PAMAM) dendrimer (Figure 1) have been studied widely as non-viral vectors for gene delivery (de Smedt et al., 2000). These cationic polymers can self-assemble with negatively-charged genes to form polymer/gene complexes (polyplexes) and induce detectable gene transfection efficiency *in vitro*. However, these first-generation polymeric gene vectors are not yet applied further for clinical practice, mainly due to low transfection efficiency and/or high cytotoxicity (Anwer et al., 2003; Merdan et al., 2002). In the past few years, extra- and intracellular gene delivery barriers have been identified that may seriously hamper efficient gene transfection (Nishikawa & Huang, 2001; Wiethoff & Middaugh, 2003). To overcome these barriers, on-going research works are devoted to molecular design of cationic polymers with multiple properties for circumventing the gene delivery barriers. It has been aware that the structures of polymers play an important role in gene transfection efficiency (Jeong et al., 2007).

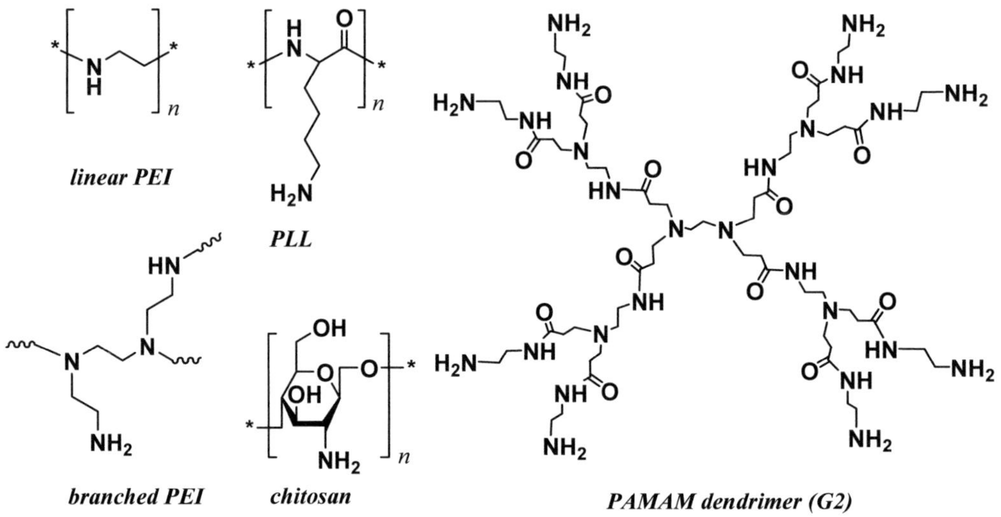

Fig. 1. Typical examples of current cationic polymers as non-viral gene delivery vectors

A lot of evidences have indicated that biodegradable cationic polymers are new-generation polymeric gene vectors because of their favourable low cytotoxicity profiles (Luten et al., 2008). Particularly, bioreducible cationic polymers, containing the disulfide bond as a bioreducible linker in polymeric main chain or side chain, are of interest. It has been well known that disulfide bond is relatively chemically stable in the extracellular environment, but can be rapidly biodegradable inside the cells due to the presence of a high amount of reducing enzymes and sulfhydryl components such as glutathione (Ganta et al., 2008; G. Wu et al., 2004). By intracellular biodegradation, smart disulfide-based cationic polymers are able to efficiently unload genes in the nucleus (Soundara & Oupicky, 2006), thereby giving rise to high levels of gene expression. Meanwhile, the biodegradation also induces relatively low cytotoxicity by avoiding the accumulation of high molecular weight cationic polymer inside cells. These efforts are now actively striving to reach safe and potent polymeric gene vectors.

In this chapter, we aim to contribute the understanding of current status on biodegradable cationic polymers for non-viral gene therapy. Fundamental knowledge on the mechanism of polymer-mediated gene delivery is described briefly. Then, first-generation polymeric gene vectors and their pros and cons are outlined. Bioreducible polymers are finally reviewed to highlight current advancement and the challenge in near feature.

2. Cationic polymer-mediated gene delivery pathway

DNA is a flexible, negatively-charged biomacromolecule under physiological conditions. It can be electrostatically repelled by negatively-charged cellular membranes and thus fails to efficiently enter the cells. Moreover, naked gene is prone to degradation by enzymes in the cells. Cationic polymers are able to condense gene into nanoscale polyplexes, deliver the genes into the cells and protect DNA from the enzymatic degradation. Therefore, for the availability of safe and potent polymeric gene delivery vectors, it is essential to understand cationic polymer-mediated gene delivery pathway.

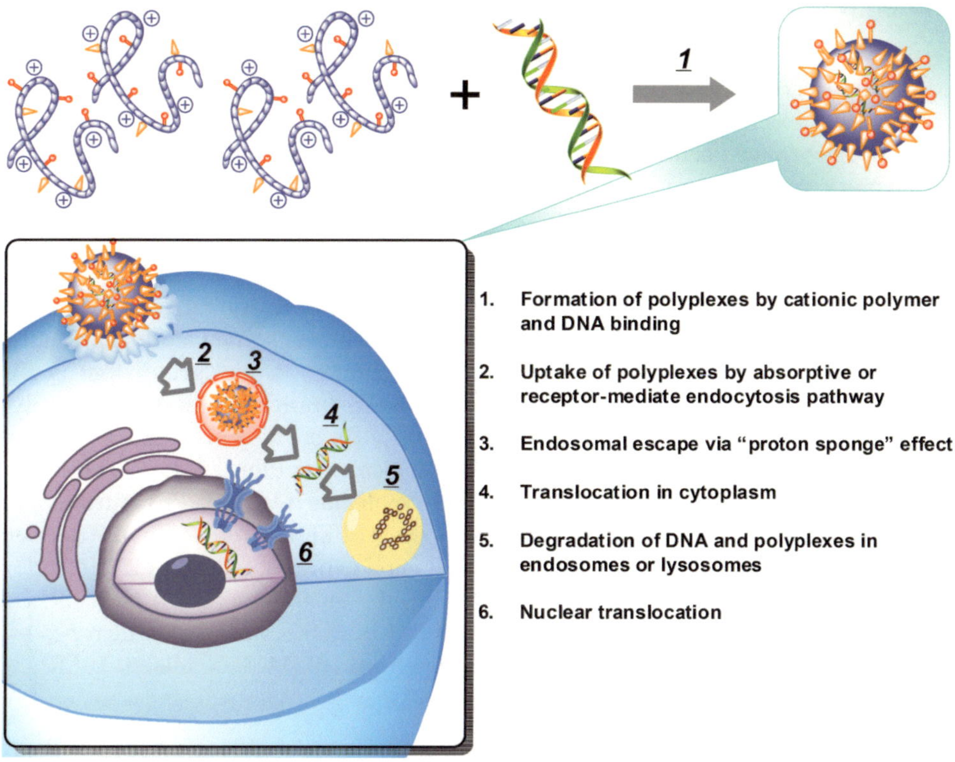

Fig. 2. Schematic illustration on cationic polymer-mediated gene delivery

A schematic gene delivery mediated by a cationic polymer is illustrated in Figure 2. First, cationic polymers bind DNA via electrostatic self-assembly to form compact polymer/DNA complexes (polyplexes). An excess amount of cationic polymer is normally needed to neutralize negative DNA and cause resulting polyplexes with net positive surface charge. Then, the positive polyplexes can interact with cellular membrane and are internalized into the cells through adsorptive endocytosis or receptor-mediated endocytosis. After they enter cells, the polyplexes normally undergo an undesirable degradation pathway from the early to later endosomes and finally locate in the lysosomes. DNA is easily degraded by enzymes in the acidic endosomes or lysosomes (pH 5~6). In this situation, cationic polymeric vectors can protect DNA from degradation and induce efficient endosomal escape by a mechanism like "proton sponge" effect (Boussif et al., 1995). After endosomal escape, polyplexes stay in the cytoplasm and move towards the nucleus by passive diffusion. At this stage, it is still unclear whether the genes should be unloaded in the cytoplasm. The polyplexes with the particle size below 25 nm may freely diffuse through the nuclear pore in the nuclear membrane (Suh et al., 2003), but the polyplexes with bigger particle size have to undergo a nuclear translocation process aided by the nuclear pore complex proteins in the nuclear membrane (Gorlich & Kutay, 1999; Ryan & Wente, 2000). When the genes are free from polyplexes in the nucleus, translation and transcription are conducted by gene expression system to produce therapeutic proteins.

3. Non-degradable cationic polymers as non-viral gene delivery vectors

This section reviews typical non-degradable cationic polymers as non-viral vectors for gene delivery. Although these polymer systems normally have low transfection capability *in vitro* and/or high cytotoxicity, from the studies on these systems, a few fundamentals on gene delivery properties have been well understood, which are valuable in the design of safe and potent polymeric gene delivery vectors.

3.1 Polyethylenimine

Polyethylenimine (PEI) was investigated as a non-viral gene delivery vector in 1995 (Boussif et al., 1995). It is a high charge density polycation, in which every three atom is present with a protonable amino-nitrogen. Linear PEI only has secondary amino group that is almost protonated under physiological conditions. By contrast, branched PEI has not only the primary and secondary amine, but the tertiary amine. As such, only about two-thirds of amino groups in PEI are protonable under physiological conditions. It has been indicated that transfection ability of PEIs depends on their molecule weights, PEI nitrogen/DNA phosphate charge ratios (N/P) and cell types. For 800-Da PEI, it can mediate the delivery of pGL2-Luc gene into NIH 3T3 cells with an optimal gene expression level of $2X10^6$ RLU/mg protein at an N/P of 8/1. However, for 25-kDa PEI, the level is increased to 10^9 RLU/mg protein at the same N/P ratio. The polyplexes of PEI may transfect different types of cell lines, with the levels of gene expression in the range from 10^5 (MCR-5 cells) to 10^8 RLU/mg protein (COS-7 cells).

Currently, high molecular weight PEI (e.g. 25kDa) is regarded as one of the most potent gene transfection agents. This superior gene transfection is explained by so-called "proton sponge" hypothesis (Boussif et al., 1995). In brief, the protonation of PEI in the endosomes induces a massive influx of chloride ions into the endosomes, which triggers the entry of water molecule into the endosome to balance the ion concentration. The entry of massive ions and water thus results in osmotic swelling of the endosome and subsequent membrane disruption. After that, genes are released into the cytoplasm. Buffer capacity (defined as the percentage of amino groups becoming protonated from pH 7.4 to 5.1) is regarded as an important parameter of cationic polymers to determine their ability to mediate endosomal escape, and is correlated with the pKa of protonable nitrogen in the polymers. Thus, cationic polymers containing protonable amino groups of a low pK_a (5-7) commonly have good buffer capacity. This may explain why branched PEI can mediate better gene transfection than linear PEI because the former has one-third of protonable tertiary amino groups.

Also, a lot of investigations on biophysical properties of PEI-based polyplexes have been made to clarify PEI-mediated gene transfection (Sirirat et al., 2003). Dynamic light scatting and zeta-potential meters are typically applied to determine the particle size and zeta-potential of polyplexes. In general, nano-scaled polyplexes below 150nm can be found with different molecular weight of PEI in the range of 2-25k at N/P ratios of 1-10. Notably, only at the N/P ratios above 4/1, the polyplexes with a high surface charge (+10~35mV) can be obtained. Small particle sizes and positive surface charges are highly desirable for efficient cellular endocytosis, which may be the reason why PEI is potent for highly efficient gene transfection.

An inherent disadvantage of PEI is its high cytotoxicity *in vitro*. Depending on cell line type, the IC_{50} value of PEI is typically below 30 μg/mL. In PEI-mediated transfection process, a two-stage cytotoxicity mechanism is discovered (Godbey et al., 2001; Moghimi et al., 2005). In the first stage, free pEI may destabilize the cellular membrane, inducing necrosis-related cytotoxicity. The removal of free PEI from the polyplexes of PEI indeed can lead to lower cytotoxicity (Boeckle et al., 2004). In the second stage, free PEI that is dissociated from the polyplexes inside the cells can interact with negatively-charged mitochondrial membrane, inducing harmful cellular apoptosis. Thus, the cytotoxicity in this stage could be diminished after cationic polymers are intracellularly degraded into small pieces.

3.2 Polyethylenimine derivatives

Low molecular weight PEI (below 2 kDa) normally displays lower cytotoxicity, but inferior transfection capability as compared to high molecular weight counterparts. Klibanov *et al.* modified the primary amines of 2k-Da PEI with dodecyl or hexadecyl iodides (M. Thomas & Klibanov, 2002). The transfection efficiencies of these alkylated 2k-Da PEI are surprising. In the transfection towards COS-7 cells, dodecylated or hexadecylated 2k-Da PEI can induce a high level of gene expression in the presence of serum, that is, 5-fold higher than that of 25k-Da PEI. The cytotoxicity of these alkylated PEI is much lower as compared to 25k-Da PEI (100% *vs.* 80% cell viability).

The incorporation of poly(ethylene glycol) (PEG) into PEI may yield PEGylated PEI with reduced cytotoxicity (C.-H. Ahn et al., 2002). PEGylated PEI copolymers can be synthesized by coupling activated PEG (2000 Da) with low molecule weight PEI (600, 1200 or 1800 Da). An optimal PEI-PEG copolymer is found that has 87 units of PEI1800 and 100 units of PEG2000. Again, the copolymer can efficiently bind plasmid DNA to form nanoscale polyplexes with positively surface charge (+20~+40mV) (average diameter 120~150nm) at N/P ratios from 1/1 to 4/1. The transfection efficiency of these polyplexes towards 293T cell is 3-fold higher than that of parent PEI1800. The cytotoxicity is very low with 80% cell viability. It should be noted that PEGylation often leads to reduced transfection efficiency *in vitro*. This is because PEGylated PEI-based polyplexes have low surface charges which impair efficient cellular internalization and also efficient endosomal escape of polyplexes (Mishra et al., 2004). Thus, the molecular weight of PEG and the composition ratio between PEI and PEG must be optimized.

3.3 Poly(L-lysine)

Poly(L-lysine) (PLL) is one of mostly studied cationic polymers for non-viral gene delivery (G.Y. Wu & Wu, 1987). It is a linear polypeptide with L-lysine residues in repeat units. The commonly used PLL as a non-viral gene delivery vector is with the molecular weigh of 25.7 kDa. The transfection efficiency of PLL is much lower than that of PEI since it displays a low buffer capacity, which is not efficient for proton sponge effect. Another disadvantage of PLL is that transfection efficiency of PLL is significantly influenced by serum probably due to the rapid binding of PLL polyplexes with negatively-charged serum (C.H. Ahn et al., 2004). In PLL-mediated gene transfection against 293T cells, for example, the gene expression level is remarkably reduced by 10 times in the presence of 10% serum. The cytotoxicity profile of PLL is not satisfactory with about 60% cell viability at a tested concentration of 10 µg/mL. Thus, modification of PLL is needed to improve transfection ability and meanwhile decrease cytotoxicity.

Because the low transfection efficiency for PLL is attributed to its poor buffer capacity, Langer *et al.* introduced imidazole group (pKa ~6.5) into PLL to improve buffer capacity, thereby enhancing transfection efficiency (Putnam et al., 2001). The modified PLL was synthesized by the coupling of amino groups of PLL (34Ka) with 4-imidazoleacetic acid using an EDC/NHS activation. As expected, transfection efficiency of imidazole-modified PLL was increased with increasing amounts of imidazole groups and was much better than that of native PLL. The PLL with the highest imidazole content (86.5%) could mediate the best gene transfection, with gene expression level close to that of polyethylenimine. Low cytotoxicity is another merit for these imidazole-modified PLL (100% cell viability at 30 µg /mL). The PEGylation of PLL can also improve the transfection ability and cytotoxicity profile of PLL (C.H. Ahn et al., 2004). For example, a group of PLL-PEG multi-block copolymers were synthesized with molecular weight in the range from 32k to 65kDa. An optimal copolymer, PLL26-co-PEG32, was found highly efficient to transfect 293T cells. The low cytotoxicity of these copolymers was also observed with more than 95% cell viability. The PLL26-co-PEG32 copolymer could mediate almost the same transfection efficiency both in the absence and presence of 10% serum.

3.4 Poly(amido amine) (PAMAM) dendrimer

Poly(amido amine) dendrimers are a family of well-defined cationic polymers (Tomalia et al., 1990). Szoka *et al.* firstly investigated PAMAM cascade polymers as non-viral gene delivery vectors (Szoka, 1993). These polymers have an ammonia initiator core and different generation (G2-G10) of amido amine repeat units. In the transfection against CV-1 cells, an optimal level of gene expression ($1x10^{10}$ RLU/mg protein) was observed for the sixth generation PAMAM (G6, MW 43451) at an N/P ratio of 6/1, but with 64% cell viability. However, 10^8 RLU/mg protein could be obtained for the PAMAM G5 at N/P=3, 6 or 10 with more than 90% cell viability. This high level of gene expression is due to a high buffer capacity of PAMAM since pKa value is 3.9 for internal tertiary amines and 6.9 for terminal amines. Another report showed that PAMAMs with an EDA initiator core can also mediate efficient gene transfection towards different mammalian cells (Kukowska-Latallo et al., 1996). The polyplexes of PAMAM G7 can transfect several types of cells with high gene transfection efficiency ($\geq 1x10^{10}$ RLU/mg protein), which is even better than that that of Lipofectamine 2000. The cytotoxicity of these PAMAMs is however terrible. It appeared that the initiator core of PAMAM may influence their transfection efficiency.

In a further study, Szoka, *et al.* found that when intact PAMAM dendrimers were treated by heating in a solvent such as water or butanol, the resulting dendrimers can surprisingly induce higher transfection efficiency as compared to parent PAMAM (Tang et al., 1993). The efficiency is affected by the generation number of dendrimer and degree of degradation. For example, after the sixth-generation dendrimer initiated with tris(2-aminoehyl)amine (TAEA), termed as 6-TAEA, was degraded in the n-butanol for 43 hours, the resulting degraded PAMAM, denoted as "fractured" dendrimer, could lead to enhanced transfection efficiency by 3 orders of magnitude compared to that of native PAMAM. This pronounced transfection efficiency is likely due to increased flexibility which results in apparent volume swelling of fractured dendrimer in the endosomes and thus efficient endosomal escape.

3.5 Chitosan

Chitosan is a naturally cationic polysaccharide and is degradable by lysozyme in the body. Chitosan is composed of β1→4 linked glucosamine, partly containing N-acetylglucosamine and has an apparent pKa value of 6.5. This indicates that chitosan could be used as non-viral vectors for gene delivery (W.G. Liu & Yao, 2002). Mumper *et al.* was the first to report chitosan as gene delivery vectors (Mumper et al., July 30-August 4, 1995). The molecular weight of chitosans influences their transfection efficiency. Generally, an optimal molecular weight is in the range of 10-50 kDa for efficient gene transfection. Also, both pH and serum largely influence the transfection efficiency of chitosan. It was shown that the transfection efficiency at pH 6.9 was higher than that at pH 7.6. Moreover, the transfection efficiency was 2 to 3 times higher in the presence of serum than that in the absence of serum. The enhancement with serum may be caused by the cell function raised by the addition of serum since major components in serum like albumin and globulin have little effect on the transfection efficiency of chitosan-based polyplexes (Sato et al., 2002).

Because chitosan is only soluble in acidic solution at pH 1~6, it readily self-aggregates under physiological conditions. A few chemical modifications on chitosan were thus performed to obtain improved solubility. One is to modify chitosan by introduction of a soluble moiety, for example, PEG, and the other is quaternization of the amine groups of chitosan (Thanou et al., 2002). A modification is also made to enhance the thermal stability of DNA by incorporating dodecylate chain into chitosan (F. Li et al., 2002). This modified chitosan displayed enhanced transfection capability with low cytotoxicity. However, the transfection efficiency of chitosan-based derivatives reported so far is normally not superior to that of PEI.

4. Hydrolysable cationic polymers as non-viral gene delivery vectors

This section reviews hydrolysable cationic polymers as non-viral gene delivery vectors. In the past decade, hydrolysable cationic polymers (Figure 3) have been studied as non-viral gene delivery vectors since they display lower toxicity as compared to their non-degradable counterparts. These research works accelerate the availability of safe and efficient non-viral gene delivery vectors.

4.1 Poly(4-hydroxy-L-proline ester) (PHP)

Poly(trans-4-hydroxy-L-proline ester) (PHP) is the first hydrolysable cationic polymers for non-viral gene delivery (Lim et al., 1999). Since hydroxyproline is a component in collagen, gelatin, and other proteins, hydroxyproline-based materials are considered low cytotoxic. PHP ester were synthesized by the polymerization of cbz-protected 4-hydroxy-L-proline to generate poly(4-hydroxy-N-cbz-L-proline), followed by the treatment of formic acid and Pd/C. As expected, PHP is degradable under physiological conditions, but very fast with degradation half time of 2 hours. Moreover, PHP can efficiently bind DNA to form positive polyplexes with average diameter below 200 nm. The polyplexes could transfect CAPE cells at a high polymer/DNA ratio of 50/1(w/w) with transfection efficiency of 1.5 times higher than that of PLL. Importantly, this polymer has very low cytotoxicity compared to 25-kDa PEI. It is worthy pointing out that the transfection of PHP is not influenced by serum, indicating this polymer is biocompatible for non-viral gene delivery.

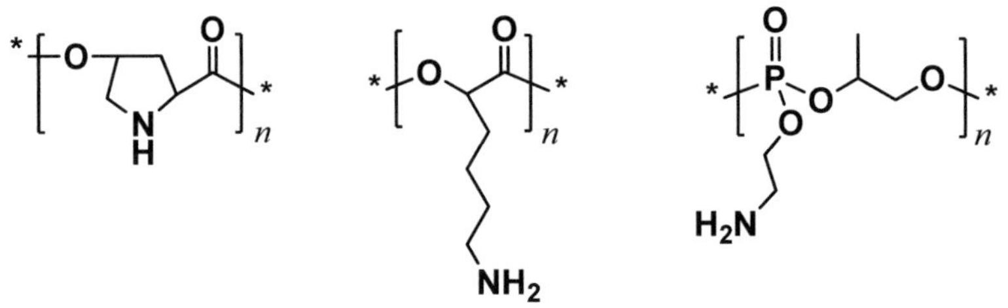

PHP *PAGA* *PPE-AE*

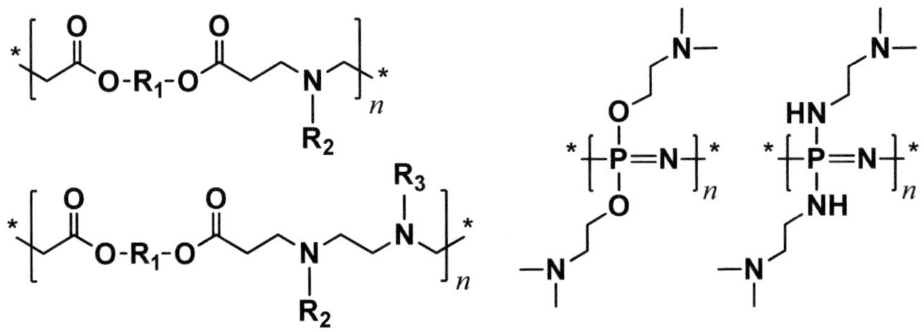

PAE *PPA*

Fig. 3. Typical examples of hydrolysable cationic polymers as non-viral vectors

4.2 Poly[α-(4-aminobutyl)-L-glycolic acid] (PAGA)

Poly[α-(4-aminobutyl)-L-glycolic acid] (PAGA, 3.3kDa) is an analogue of PLL (Lim et al., 2000). Since the amide bonds in PLL are replaced with ester bonds, PAGA is degradable under physiological conditions. PAGA is rapidly degraded at pH 7.4 and 37°C with the degradation half time of 30 min. The presence of primary amine in the side chain of PAGA renders this polymer highly efficient for gene binding. As such, at a low N/P ratio of 5/1, nanoscale polyplexes from PAGA/DNA (~326nm) can be formed [34]. The transfection efficiency of PAGA is comparable with that of PLL. At an optimal N/P of 40/1, the transfection efficiency of the polyplexes of PAGA is 3-fold higher compared to those of 4-kDa PLL. The cytotoxicity of PAGA is much low in comparison with that of PLL (100% vs. 20% cell viability). Although PAGA is not as efficient as PEI for gene transfection, the study on the PAGA indicates that biodegradable cationic polymers are relatively safer due to low cytotoxicity as compared to non-degradable cationic polymers.

4.3 Poly(β-aminoester)s (PAE)

Poly(β-amino ester)s (PAEs) are a family of degradable cationic polymers that are prepared via Michael-type addition between bisacrylates and amines. Since these reactive monomers are versatile, a large number of PAEs with different functional groups can be designed (Lynn & Langer, 2000). Due to the presence of multiple ester bonds in polymer main chain, PAEs are degradable and normally possess relatively low cytotoxicity. A particular example is that Langer et al. examined a library containing more than 2000 PAEs for non-viral gene delivery (Anderson et al., 2003). Also, the structure-activity relationships were investigated. The conclusions from the study are: 1) Bisacrylate monomers with strongly hydrophobic residues are almost always present in the 50 best-performing PAEs; 2) Linear, bis(secondary amines) are over represented in the hit structures; 3) Mono- or dialcohol side group in PAE is an important functional entity for efficient gene transfection. One PAE from this library showed transfection ability, with the level of gene expression 5-fold higher than that of 25- kDa PEI against 3T3 cell lines under optimal conditions. The studies on PAEs strongly support the idea that degradable cationic polymers are very promising for safe and efficient gene delivery.

4.4 Polyphosphoester (PPE-EA)

Poly(2-amioethyl propylene phosphate) (PPE-EA) is a degradable cationic polymer which can yield ultimate low-toxic degradation products including α-propylene glycol, phosphate, and ethanolamine. This polymer is synthesized through ring-opening polymerization of 4-methyl-2-oxo-2-hydro-1,3,2-dioxaphospholane, followed by two-step chemical modification (Wang et al., 2001). The transfection of PPE-EA (30 kDa) gives 100-fold higher levels of gene expression at an N/P ratio of 6 as compared to that of 27- kDa PLL at an N/P ratio of 5. Importantly, PPE-EA has low cytotoxicity towards COS-7 cells with more than 80% cell viability up to a tested concentration of 1000 μg/mL. However, this polymer has poor ability of endosomal escape since the presence of chloroquine (100 μM), a reagent known to disrupt endosomal membrane, may lead to remarkably enhanced transfection efficiency.

4.5 Polyphosphazene (PPA)

Polyphosphazenes (PPAs) are cationic polymers derived from poly(dichloro)phosphazene (Luten et al., 2003). They are degradable slowly under physiological conditions with a half-life of more than 10 days, but relatively faster at pH 5 with a half-life of 4-5 days. The PPAs (>100 kDa) can condense genes into nanoscale polyplexes at low polymer/DNA mass ratios of 3~5, with a high positive surface charge (+ 40mV). The transfection of PPA was comparable with that of PEI towards COS-7 cells. The optimal transfection efficiency of PPA was observed at a polymer/DNA mass ratio of 6, however, with a high cytotoxicity profile (~50% cell viability). The pronounced cytotoxicity could be due to high molecular weight and slow degradation profile inside the cells.

5. Bioreducible cationic polymers as non-viral gene delivery vectors

In the design of hydrolysable cationic polymers for non-viral gene delivery, a contradiction has to be found that the polymers are expected to be rapidly degradable intracellularly as one hand, but chemically stable extracellularly as another hand. In order to avoid this issue, disulfide bond as a bioreducible linker has received much attention in recent years. The disulfide bond is chemically stable in the blood plasma, but intracellularly bio-cleavable by reducing enzymes like glutathione reductase and sulfhydryl components like glutathione since the concentration of these reducing species is much higher in the cytoplasm than the blood plasma (intracellular vs. extracellular glutathione concentration, 0.5-10 mM vs. 2-20 μM) (G. Wu et al., 2004). Thus, this feature makes the disulfide very valuable in the design of biodegradable cationic polymers for triggered gene delivery. Figure 4 shows a schematic illustration on intracellular gene delivery mediated by disulfide-based cationic polymers.

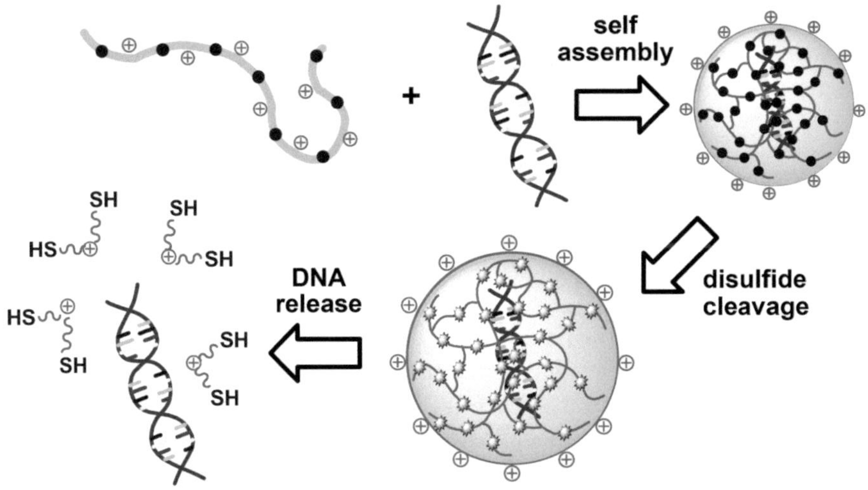

Fig. 4. A conceptual illustration of DNA binding and subsequent intracellular release: (a) formation of the polyplexes of bioreducible cationic polymers, which are relatively stable in the extracellular environment, (b) intracellular cleavage of disulfide linkages in the polymer of the polyplex, and (c) intracellular DNA release from the degraded polymer.

This section reviews current progress in disulfide-based cationic polymers as non-viral gene delivery vectors. The topics are focused on the synthesis of bioreducible cationic polymers, unique biophysical properties of the polyplexes based on the polymers.

5.1 Preparation of bioreducible cationic polymers as non-viral gene vectors

Bioreducible cationic polymers can be designed and synthesized that contain disulfide bond either in polymer main chain or side chain. In earlier studies, the disulfide was introduced in the polymer side chain to conceptually confirm the role of the disulfide in gene delivery. A typical synthesis route is the preparation of cationic polymers with pyridyldithio residue, which is then modified with suitable thiol compounds via an exchange reaction (Figure 5a). By this method, the pLL containing disulfide linkages in the polymer side chains (termed as poly[Lys-(AEDTP)]) was prepared through chemical modification of the primary amines in pLL with N-succinimidyl-3-(2-pyridyldithio)propionate, followed by an exchange reaction with mercapthoethylamine (Pichon et al., 2002). The polyplexes of poly[Lys-(AEDTP)] can transfect HeLa cells with a level of gene expression 10-fold higher than that of parent pLL. This thus implies that disulfide linker plays a pivotal role in improved gene transfection. In another work, PAEs with pyridyldithio groups in the polymer side chains were synthesized via Michael-type addition reaction between diacrylates and 2-(pyridyldithio)-ethylamine. These polymers were further modified with mercaptoethylamine or thiol peptide such as RGD, yielding the PAEs with disulfide linkers in the side chains (SS-PAEs) (Zugates et al., 2006). The polyplexes of SS-PAEs could transfect HCC cells with the efficiency comparable to that of 25-kDa PEI.

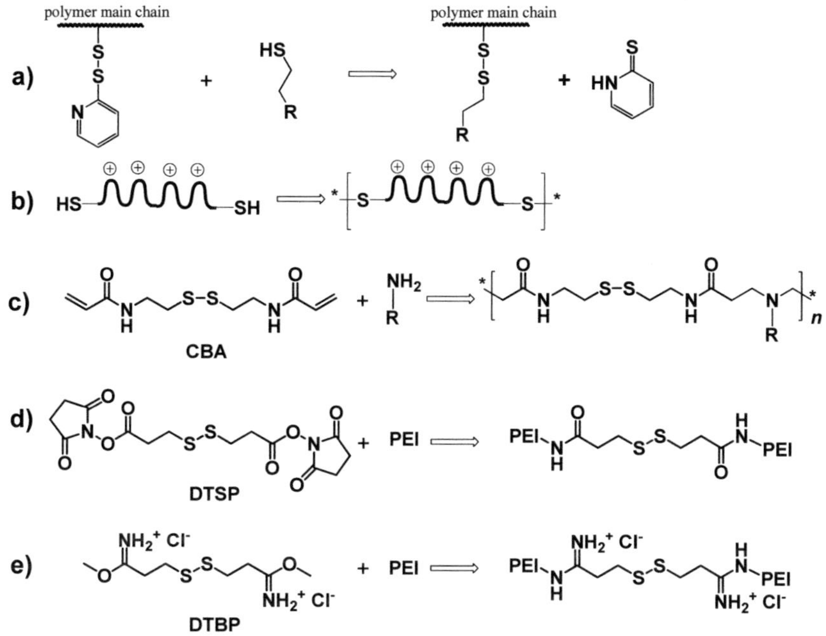

Fig. 5. Typical methods for the preparation of bioreducible cationic polymers as non-viral gene delivery vectors

Alternatively, one route to generate bioreducible cationic polymers is the polyoxidation of dithiol-based monomers having amino groups (Figure 5b). Typical examples are disulfide-containing cationic polymers based on pEI , pLL and pDMAEMA (SS-PEI, SS-PLL and SS-PDMAEMA, respectively, in Figure 5). In general, the preparation of these dithiol-based oligoamines is time-consuming and these compounds can not be stored for long term due to oxidation of thiol groups by air. As typical examples, Park *et al.* reported on the synthesis of dithiol-containing oligoamines via organic synthesis involving protection and deprotection of amino groups (Lee et al., 2007). Oupický *et al.* described the preparation of well-defined dithiol-based PDMAEMA oligomers via reversible addition-fragmentation chain transfer polymerization (You et al., 2007). Seymour *et al.* produced dithiol-based oligopeptides (Cys-Lys10-Cys) via solid-phase organic synthesis (Oupicky et al., 2002). These dithiol-based oligoamines can be oxidized by DMSO as an oxidant agent to yield disulfide-containing cationic polymers. Also, different dithiol-bearing groups, e.g. nuclear localization sequences comprising two cysteine residues, can be incorporated in the oxidation reaction, giving rise to disulfide-containing copolymers with multiple functionalities (Read et al., 2005).

A simple approach for the availability of disulfide-based cationic polymers is the chemical coupling of amine compounds with disulfide-containing reagents, such as cystamine bisacrylamide (CBA) in a Michael addition reaction (Lin et al., 2006, 2007a; Lin et al., 2007b; Lin et al., 2008; Lin & Engbersen, 2008) (Figure 5c), and dithiobis(succinimidyl propionate) (DTSP) or dithiobispropionimidate (DTBP) in a polycondensation reaction (Figures 5d&e). These reactions can generate linear or branched disulfide-containing cationic polymers with different molecular structures (Figure 6). Lee *et al.* firstly prepared disulfide-containing branched pEI by the crosslinking of low molecular weight PEI with DTSP or DTBP (Gosselin et al., 2001; Gosselin et al., 2002). Recently, disulfide-containing poly(amido amine) (SS-PAA) (co)polymers were synthesized through Michael-type addition reaction of CBA to primary amines, secondary diamines or PEI oligoamines. The structural effects of these SS-PAAs on gene delivery properties were systematically investigated. It was shown that the SS-PAA with the hydroxybutyl or hydroxypentyl side groups led to higher transfection efficiencies and lower cytotoxicity in COS-7 cells than 25-kDa branched PEI. Herein, we summarize those typical bioreducible cationic polymers in Table 1 and their performance in gene transfection efficiency against different cell lines.

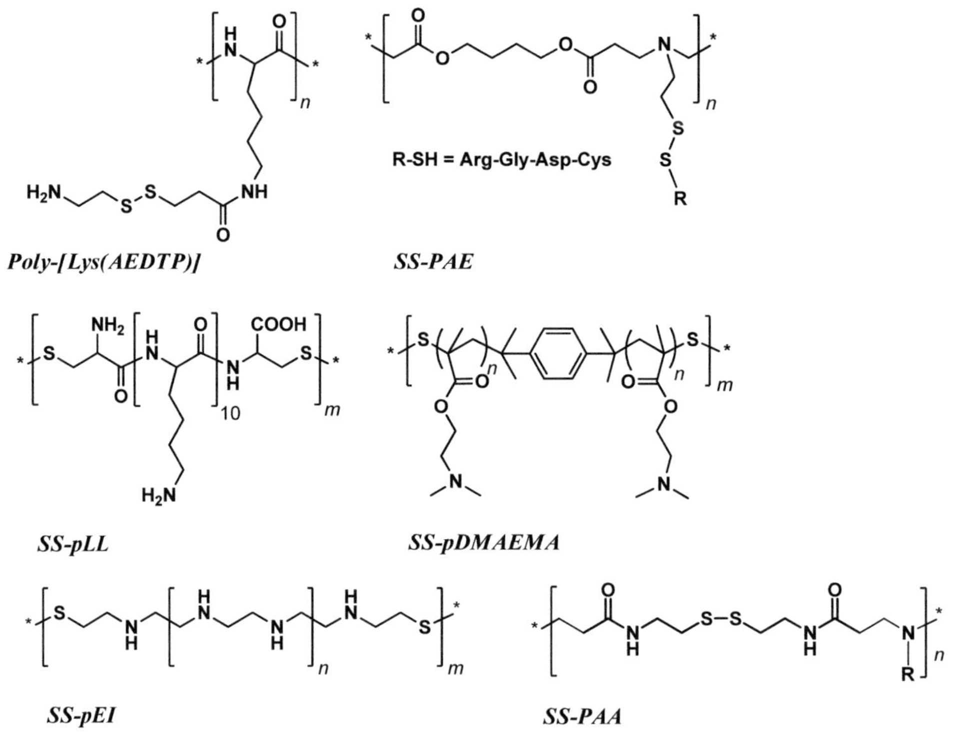

Fig. 6. Typical examples of bioreducible cationic polymers as non-viral gene delivery vectors

5.2 Intracellular fate of disulfide-based polymeric gene vectors

It is assumed that higher efficient transfection induced by disulfide-based cationic polymers is at least partly due to intracellular degradation via the cleavage of the disulfides in the reducing intracellular environment. In order to obtain experimental evidences, polyplexes of fluorescently labelled P(CBA-ABOL) containing disulfide bonds and P(BAPABOL) lacking disulfide bonds (Figure 7) were used for gene transfection against COS-7 cells at the same polymer/DNA mass ratio of 12/1. Dynamic light scattering and zeta-potential measurement showed that polyplexes of P(CBA-ABOL) and p(BAP-ABOL) had comparable average particle size and surface charge (128 nm vs. 82 nm; +20.2 mV vs. +19.2 mV), allowing good comparison of the transfection activity of both types of polyplexes. The intracellular distributions of the two polymers, labelled by a Rhodamine dye, are clearly different under fluorescence microscopy (Figure 7). For

P(CBA-ABOL), a homogeneous dispersed fluorescence was observed both in the cytoplasm and the nucleus. By contrast, for the p(BAP-ABOL) lacking the disulfide linkages, many micro-sized aggregated clumps were found in the perinuclear space and only a few weak fluorescence was observed in the nucleus. These results may serve as an indication that P(CBA-ABOL) is intracellularly degradable faster by reducing cleavage of the disulfide bonds, resulting in a diffuse distribution of fluorescently labeled polymer fragments inside the cells. Slow degradation of p(BAP-ABOL) may, however, contributes to the formation of the aggregation.

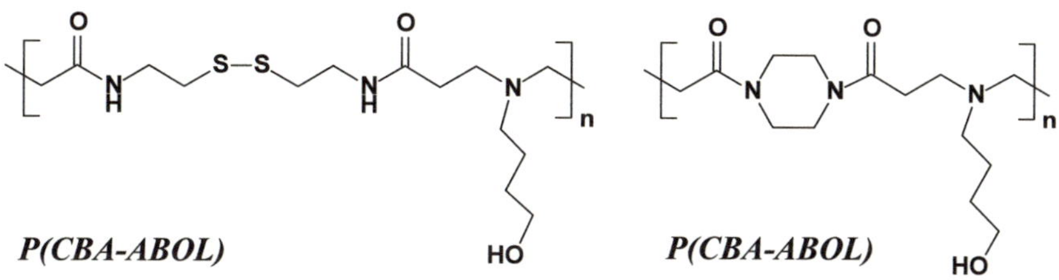

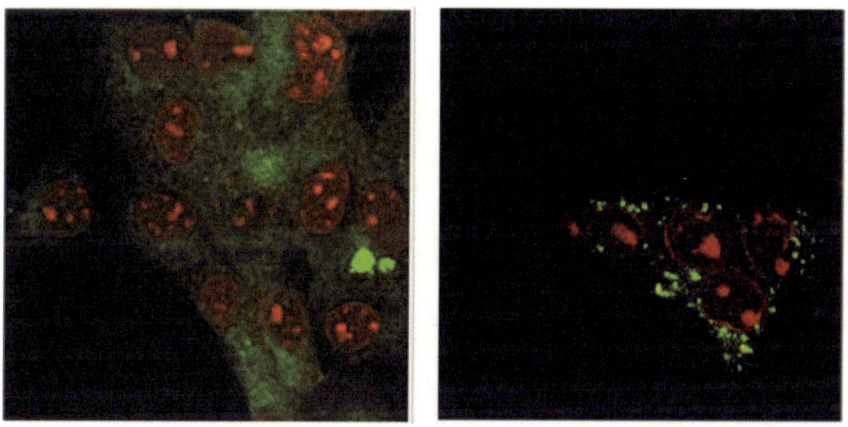

Fig. 7. Intracellular distribution of the polyplexes from bioreducible P(CBA-AOBL) (left) and non-degradable P(BAP-ABOL) lacking disulfide bonds (right), observed under confocal laser scanning microscopy. The polymers are shown in green and the nucleus in red.

Disulfide-based polymers	Plasmid	Cell line/animal	Transfection	Ref.
DTSP or DTBP-crosslinked PEI	pCMV-Luc	CHO	Comparable to 25-kDa PEI	(Gosselin et al., 2001)
PEGylated PEI crosslinked with DTSP	pCMV-Luc	Mice	enhanced plasmid blood levels up to 60 min.	(Neu et al., 2007a)
Crosslinked PEI with DSP	pCMV-Luc, ^{32}P-labeled plasmid	NIH-3T3/ Balb/c mouse	3-fold higher than 25-kDa PEI, considerable gene expression in the liver and lung	(Neu et al., 2007b)

Disulfide-based polymers	Plasmid	Cell line/animal	Transfection	Ref.
Linear SS-PEI	pCMV-Luc	HepG2, HeLa	comparable to ExGen 500 (25kDa *l*-PEI)	(Lee et al., 2007)
DTSP-Crosslinked linear PEI (2-4kDa)	pCMV-EGFP	CHO, HepG2, NIH-3T3, HeLa, HCT116, COS-7, HEK293	5-7 times higher than Lipofectamine 2000, JetPEI, FuGENE6, 40-70 eGFP$^+$%	(Breunig et al., 2007)
PEI-SS(x) from thiolated 800-kDa	pCMV-Luc, pCMV-EGFP	HeLa, 293T	10-fold higher than 25- kDa PEI (HeLa), 3-fold higher eGFP$^+$% (293T)	(Peng et al., 2008)
listeriolysinO-conjugated reducible 25kDa-PEI (LLO-SS-PEI)	pCMV-NGVL3 (both GFP and Luc)	HEK293	Comparable to 25-kDa PEI	(Choi & Lee, 2008)
SS-PEI in the presence of RGD	pCMV-Luc and pEGFP	293T, HeLa	Comparable (293T), 8-fold higher (HeLa) than 25 kDa PEI	(Sun et al., 2008)
BPEI-SS-PEG-cNGR (cNGR: cyclic NRG (CNGRCK) peptide)	Luciferase gene pDNA	HEK293, HT1080	100-fold higher than that of 1.2kDa-BPEI	(Son et al., 2010)
branched poly(ethylenimine sulfide) (b-PEIS)	pCN-Luc or pEGFP	HEK293, HeLa, NIH3T3, C2C12, HUVECs	1000-fold higher than 6-kDa *l*-PEI, but comparable to 25-kDa BPEI	(Koo et al., 2010)
CBA-crosslinked reductable polyspermine	pEGFP	A549	4-fold higher than 25-kDa PEI	(Jere et al., 2009)
Reducible PEI (PEI-SS-CLs) via "click" chemistry	pLuc, pEGFP	293T, HeLa	5-10 times higher than 25-kDa PEI	(J. Liu et al., 2010)
Linear PAA grafted with polyamidoamines	pLuc, pEGFP	293T, HeLa	Comparable (293T) or a little higher (HeLa) than 25-kDa BPEI	(Xue et al., 2010)
CBA-crosslined reducible polyamines (pLPEI/pTETA/pSPE)	qWIZ-Luc (6.7 kb), qWIZ-GFP (5.7 kb)	murine brain capillary endothelial bEnd.3 cells	2.3-4.9 fold higher than ExGen500	(Zhang & Vinogradov, 2010)
linear disulfide-based "click" polymer (RCP)	iMDR1-pDNA, pEGFP	MCF-7, MCF-7/ADR	Comparable to 25kDa- BPEI	(Gao et al., 2010)

Table 1. A summary of bioreducible cationic polymers and their transfection efficiencies in different types of cell lines.

It appears that disulfide degradation mainly proceeds in the cytoplasm and in the nucleus. However, a few recent studies showed that, depending on the cell line type and the polymer constructs, the disulfide could also be degradable in those microenvironments such as the cellular surface, the endosomes and the lysosomes (Blacklock et al., 2009; Morre & Morre, 2003). Thus, further studies are certainly needed to understand the factors influencing the degradation at specific locations.

6. Conclusion

Cationic polymers with multiple functionalities are promising as non-viral vectors for gene transfection. Since more and more extracellular and intracellular gene delivery barriers are identified that seriously hamper efficient gene transfection, a number of cationic polymers have been designed that are capable of overcoming one or more gene delivery barriers, thus leading to detectable gene transfection efficiency. From those conventional non-degradable cationic polymers to current bioreducible cationic polymers, peoples have more and more reached virus-like, safe and potent polymeric gene delivery vectors. Further understanding on structure-activity relationships of cationic polymers and their intracellular fate should be indispensable, in order to achieve polymer systems that can exhibit multiple gene delivery properties for highly efficient gene transfection.

7. Acknowledgment

This work was financially supported by the Innovation Program of Shanghai Municipal Education Commission (No. 10ZZ26), the Program for Young Excellent Talents in Tongji University (No. 2009KJ077), the Scientific Research Foundation for the Returned Overseas Chinese Scholars, State Education Ministry, and the National Natural Science Foundation of China (No. 20904041).

8. References

Ahn, C.-H., Chae, S.Y., Bae, Y.H. & Kim, S.W. (2002). Biodegradable Poly(Ethylenimine) for Plasmid DNA Delivery. *Journal of Controlled Release,* Vol.80, No.-3, (April 2002), pp. 273-282, ISSN 0168-3659

Ahn, C.H., Chae, S.Y., Bae, Y.H. & Kim, S.W. (2004). Synthesis of Biodegradable Multi-Block Copolymers of Poly(L-Lysine) and Poly(Ethylene Glycol) as a Non-Viral Gene Carrier. *Journal of Controlled Release,* Vol.97, No.3, (July 2004), pp. 567-574, ISSN 0168-3659

Anderson, D.G., Lynn, D.M. & Langer, R. (2003). Semi-Automated Synthesis and Screening of a Large Library of Degradable Cationic Polymers for Gene Delivery. *Angewandte Chemie International Edition,* Vol.42, No.27, (July 2003), pp. 3153-3158, ISSN 1433-7851

Anwer, K., Rhee, B.G. & Mendiratta, S.K. (2003). Recent Progress in Polymeric Gene Delivery Systems. *Critical Reviews in Therapeutic Drug Carrier Systems,* Vol.20, No.4, (July-August 2003), pp. 249-293, ISSN 0743-4863

Blacklock, J., You, Y.-Z., Zhou, Q.-H., Mao, G. & Oupický, D. (2009). Gene Delivery in Vitro and in Vivo from Bioreducible Multilayered Polyelectrolyte Films of Plasmid DNA. *Biomaterials,* Vol.30, No.5, (February 2009), pp. 939-950, ISSN 0142-9612

Boeckle, S., von Gersdorff, K., van der Piepen, S., Culmsee, C., Wagner, E. & Ogris, M. (2004). Purification of Polyethylenimine Polyplexes Highlights the Role of Free

Polycations in Gene Transfer. *Journal of Gene Medicine,* Vol.6, No.10, (October 2004), pp. 1102-1111, ISSN 1099-498X

Boussif, O., Lezoualch, F., Zanta, M.A., Mergny, M.D., Scherman, D., Demeneix, B. & Behr, J.P. (1995). A Versatile Vector for Gene and Oligonucleotide Transfer into Cells in Culture and in Vivo: Polyethylenimine. *Proceedings of the National Academy of Sciences of the United States of America,* Vol.92, No.16, (August 1995), pp. 7297-7301, ISSN 0027-8424

Breunig, M., Lungwitz, U., Liebl, R. & Goepferich, A. (2007). Breaking up the Correlation between Efficacy and Toxicity for Nonviral Gene Delivery. *Proceedings of the National Academy of Sciences of the United States of America,* Vol.104, No.36, (September 2007), pp. 14454-14459, ISSN 0027-8424

Choi, S. & Lee, K.D. (2008). Enhanced Gene Delivery Using Disulfide-Crosslinked Low Molecular Weight Polyethylenimine with Listeriolysin O-Polyethylenimine Disulfide Conjugate. *Journal of Controlled Release,* Vol.131, No.1, (October 2008), pp. 70-76, ISSN 0168-3659

de Smedt, S.C., Demeester, J. & Hennink, W.E. (2000). Cationic Polymer Based Gene Delivery Systems. *Pharmaceutical Research,* Vol.17, No.2, (February 2000), pp. 113-126, ISSN 1043-6618

El-Aneed, A. (2004). An Overview of Current Delivery Systems in Cancer Gene Therapy. *J Control Rel,* Vol.94, No.2004), pp. 1-14,

Ganta, S., Devalapally, H., Shahiwala, A. & Amiji, M. (2008). A Review of Stimuli-Responsive Nanocarriers for Drug and Gene Delivery. *Journal of Controlled Release,* Vol.126, No.3, (March 2008), pp. 187-204, ISSN 0168-3659

Gao, Y., Chen, L., Zhang, Z., Chen, Y. & Li, Y. (2010). Reversal of Multidrug Resistance by Reduction-Sensitive Linear Cationic Click Polymer/iMDR1-pDNA Complex Nanoparticles. *Biomaterials,* Vol.32, No.6, (February 2010), pp. 1738-1747, ISSN 0142-9612

Godbey, W.T., Wu, K.K. & Mikos, A.G. (2001). Poly(Ethylenimine)-Mediated Gene Delivery Affects Endothelial Cell Function and Viability. *Biomaterials,* Vol.22, No.5, (March 2001), pp. 471-480, ISSN 0142-9612

Gorlich, D. & Kutay, U. (1999). Transport between the Cell Nucleus and the Cytoplasm. *Annual Review of Cell and Developmental Biology,* Vol.15, No.1, (January 1999), pp. 607-660, ISSN 1081-0706

Gosselin, M.A., Guo, W. & Lee, R.J. (2001). Efficient Gene Transfer Using Reversibly Cross-Linked Low Molecular Weight Polyethylenimine. *Bioconjugate Chemistry,* Vol.12, No.6, (November-December 2001), pp. 989-994, ISSN 1043-1802

Gosselin, M.A., Guo, W. & Lee, R.J. (2002). Incorporation of Reversibly Cross-Linked Polyplexes into LPDII Vectors for Gene Delivery. *Bioconjugate Chemistry,* Vol.13, No.5, (September-October 2002), pp. 1044-1053, ISSN 1043-1802

Jeong, J.H., Kim, S.W. & Park, T.G. (2007). Molecular Design of Functional Polymers for Gene Therapy. *Progress in Polymer Science,* Vol.32 No.11, (November 2007), pp. 1239-1274, ISSN 0079-6700

Jere, D., Kim, J.E., Arote, R., Jiang, H.L., Kim, Y.K., Choi, Y.J., Yun, C.H., Cho, M.H. & Cho, C.S. (2009). Akt1 Silencing Efficiencies in Lung Cancer Cells by Sh/Si/Ssirna Transfection Using a Reductable Polyspermine Carrier. *Biomaterials,* Vol.30, No.8, (March 2009), pp. 1635-1647, ISSN 0142-9612

Koo, H., Jin, G., Kang, H., Lee, Y., Nam, K., Zhe, B.C. & Park, J.S. (2010). Biodegradable Branched Poly(Ethylenimine Sulfide) for Gene Delivery. *Biomaterials,* Vol.31, No.5, (February 2010), pp. 988-997, ISSN 0142-9612

Kukowska-Latallo, J.F., Bielinska, A.U. & Johnson, J. (1996). Efficient Transfer of Genetic Material into Mammalian Cells Using Starburst Polyamioamine Dendrimers. *Proceedings of the National Academy of Sciences of the United States of America,* Vol.93 No.10, (May 1996), pp. 4897-4902, ISSN 0027-8424

Lee, Y., Mo, H., Koo, H., Park, J.Y., Cho, M.Y., Jin, G.W. & Park, J.S. (2007). Visualization of the Degradation of a Disulfide Polymer, Linear Poly(Ethylenimine Sulfide), for Gene Delivery. *Bioconjugate Chemistry,* Vol.18 No.1, (January-February 2007), pp. 13-18, ISSN 1043-1802

Li, F., Liu, W.G. & Yao, K.D. (2002). Preparation of Oxidized Glucose-Crosslinked N-Alkylated Chitosan Membrane and in Vitro Studies of pH-Sensitive Drug Delivery Behavior. *Biomaterials,* Vol.23, No.2, (January 2002), pp. 343-347, ISSN 0142-9612

Li, S. & Huang, L. (2000). Non-Viral Gene Therapy: Promises and Challenges. *Gene Therapy,* Vol.7, No.1, (January 2000), pp. 31-34, ISSN 0969-7128

Lim, Y.B., Choi, Y.H. & Park, J.S. (1999). A Self-Destroying Polycationic Polymer: Biodegradable Poly(4-Hydroxy-L-Proline Ester). *Journal of the American Chemical Society,* Vol.121, No.24, (June 1999), pp. 5633-5639, ISSN 0002-7863

Lim, Y.B., Kim, C.H., Kim, K., Kim, S.W. & Park, J.S. (2000). Development of a Safe Gene Delivery System Using Biodegradable Polymer, Poly[Alpha-(4-Aminobutyl)-L-Glycolic Acid]. *Journal of the American Chemical Society,* Vol.122, No.4, (June 2000), pp. 6524-6525, ISSN 0002-7863

Lin, C., Zhong, Z.Y., Lok, M.C., Jiang, X., Hennink, W.E., Feijen, J. & Engbersen, J.F.J. (2006). Linear Poly(Amido Amine)S with Secondary and Tertiary Amino Groups and Variable Amounts of Disulfide Linkages: Synthesis and in Vitro Gene Transfer Properties. *Journal of Controlled Release,* Vol.116, No.2, (November 2006), pp. 130-137, ISSN 0168-3659

Lin, C., Zhong, Z.Y., Lok, M.C., Jiang, X., Hennink, W.E., Feijen, J. & Engbersen, J.F.J. (2007a). Random and Block Copolymers of Bioreducible Poly(Amido Amine)S with High- and Low-Basicity Amino Groups: Study of DNA Condensation and Buffer Capacity on Gene Transfection. *Journal of Controlled Release,* Vol.123, No.1, (October 2007a), pp. 67-75, ISSN 0168-3659

Lin, C., Zhong, Z.Y., Lok, M.C., Xulin Jiang, Wim E. Hennink, Feijen, J. & Engbersen, J.F.J. (2007b). Novel Bioreducible Poly(Amido Amine)S for Highly Efficient Gene Delivery. *Bioconjugate Chemistry,* Vol.18, No.1, (January-February 2007b), pp. 138-145, ISSN 1043-1802

Lin, C., Blaauboer, C.-J., Timoneda, M.M., Lok, M.C., van Steenbergen, M., Hennink, W.E., Feijen, J., Zhong, Z.Y. & Engbersen, J.F.J. (2008). Bioreducible Poly(Amido Amine)S with Oligoamine Side Chains: Synthesis, Characterization, and Structural Effect on Gene Delivery. *Journal of Controlled Release,* Vol.126, No.2, (March 2008), pp. 166-174, ISSN 0168-3659

Lin, C. & Engbersen, J.F.J. (2008). Effects of Chemical Functionalities in Poly(Amido Amine)S for Non-Viral Gene Transfection. *Journal of Controlled Release,* Vol.132, No.3, (December 2008), pp. 267-272, ISSN 0168-3659

Liu, J., Jiang, X., Xu, L., Wang, X., Hennink, W.E. & Zhuo, R. (2010). Novel Reduction-Responsive Cross-Linked Polyethylenimine Derivatives by Click Chemistry for

Nonviral Gene Delivery. *Bioconjugate Chemistry,* Vol.21, No.10, (October 2010), pp. 1827-1835, ISSN 1043-1802

Liu, W.G. & Yao, K.D. (2002). Chitosan and Its Derivatives-a Promising Non- Viral Vector for Gene Transfection. *Journal of Controlled Release,* Vol.83, No.1, (September 2002), pp. 1-11, ISSN 0168-3659

Luo, D. & Saltzman, W.M. (2000). Synthetic DNA Delivery Systems. *Nature Biotechnology,* Vol.18, No.1, (January 2000), pp. 33-37, ISSN 1087-0156

Luten, J., van Steenis, J.H. & van Someren, R. (2003). Water-Soluble Biodegradable Cationic Polyphosphazenes for Gene Delivery. *Journal of Controlled Release,* Vol.89, No.3, (May 2003), pp. 483-497, ISSN 0168-3659

Luten, J., van Nostruin, C.F., De Smedt, S.C. & Hennink, W.E. (2008). Biodegradable Polymers as Non-Viral Carriers for Plasmid DNA Delivery. *Journal of Controlled Release,* Vol.126, No.2, (March 2008), pp. 97-110, ISSN 0168-3659

Lynn, D.M. & Langer, R. (2000). Degradable Poly(B-Amino Esters): Synthesis, Characterization, and Self-Assembly with Plasmid DNA. *Journal of the American Chemical Society,* Vol.122, No.44, (October 2000), pp. 10761-10768, ISSN 0002-7863

Merdan, T., Kopecek, J. & Kissel, T. (2002). Prospects for Cationic Polymers in Gene and Oligonucleotide Therapy against Cancer. *Advanced Drug Delivery Review,* Vol.54, No.5, (September 2002), pp. 715-758, ISSN 0169-409X

Mishra, S., Webster, P. & Davis, M.E. (2004). Pegylation Significantly Affects Cellular Uptake and Intracellular Trafficking of Non-Viral Gene Delivery Particles. *European Journal of Cell Biology,* Vol.83, No.3, (April 2004), pp. 97-111, ISSN 0171-9335

Moghimi, S.M., Symonds, P., Murray, J.C., Hunter, A.C., Dekska, G. & Szewczyk, A. (2005). A Two-Stage Poly(Ethylenimine)-Mediated Cytotoxicity: Implications for Gene Transfer/Therapy. *Molecular Therapy,* Vol.11, No.6, (June 2005), pp. 990-995, ISSN 1525-0016

Morre, D.J. & Morre, D.M. (2003). Cell Surface NADH Oxidases (ECTO-NOX Proteins) with Roles in Cancer, Cellular Time-Keeping, Growth, Aging and Neurodegenerative Diseases. *Free Radical Research,* Vol.37, No.8, (August 2003), pp. 795-808, ISSN 1071-5762

Mumper, R.J., Wang, J., Claspell, J.M. & Rolland, A.P. (1995). Novel Polymeric Condensing Carriers for Gene Delivery. *Proceedings of the 21st International Symposium of Controlled Release of Bioactive Materials,* ISBN 0849351812, Seattle, Washington, July 30-August 4, 1995

Neu, M., Germershaus, O., Behe, M. & Kissel, T. (2007a). Bioreversibly Crosslinked Polyplexes of PEI and High Molecular Weight PEG Show Extended Circulation Times in Vivo. *Journal of Controlled Release,* Vol.124, No.1-2, (December 2007a), pp. 69-80, ISSN 0168-3659

Neu, M., Germershaus, O., Mao, S., Voigt, K., Behe, M. & Kissel, T. (2007b). Crosslinked Nanocarriers Based Upon Poly(Ethylene Imine) for Systemic Plasmid Delivery: In Vitro Characterization and in Vivo Studies in Mice. *Journal of Controlled Release,* Vol.118, No.3, (April 2007b), pp. 370-380, ISSN 0168-3659

Nishikawa, M. & Huang, L. (2001). Nonviral Vectors in the New Millennium: Delivery Barriers in Gene Transfer. *Nucleic Acid Research,* Vol.12 No.8, (May 2001), pp. 861-870, ISSN 0305-1048

Oupicky, D., Parker, A.L. & Seymour, L.W. (2002). Laterally Stabilized Complexes of DNA with Linear Reducible Polycations: Strategy for Triggered Intracellular Activation of DNA Delivery Vectors. *Journal of the American Chemical Society,* Vol.124, No.1, (January 2002), pp. 8-9, ISSN 0002-7863

Pack, D.W., Hoffman, A.S., Pun, S. & Stayton, P.S. (2005). Design and Development of Polymers for Gene Delivery. *Nature Reviews Drug Discovery,* Vol.4, No.7, (July 2005), pp. 589-593, ISSN 1474-1776

Peng, Q., Zhong, Z.L. & Zhuo, R.X. (2008). Disulfide Cross-Linked Polyethylenimines (PEI) Prepared Via Thiolation of Low Molecular Weight PEI as Highly Efficient Gene Vectors. *Bioconjugate Chemistry,* Vol.19, No.2, (February 2008), pp. 499-506, ISSN 1043-1802

Pichon, C., LeCam, E., Guerin, B., Coulaud, D., Delain, E. & Midoux, P. (2002). Poly [Lys-(AEDTP)]: A Cationic Polymer That Allows Dissociation of Pdna/Cationic Polymer Complexes in a Reductive Medium and Enhances Polyfection. *Bioconjugate Chemistry,* Vol.13, No.1, (January-February 2002), pp. 76-82, ISSN 1043-1802

Putnam, D., Gentry, C.A., Pack, D.W. & Langer, R. (2001). Polymer-Based Gene Delivery with Low Cytotoxicity by a Unique Balance of Side-Chain Termini. *Proceedings of the National Academy of Sciences of the United States of America,* Vol.98, No.3, (January 2001), pp. 1200-1205, ISSN 0027-8424

Read, M.L., Singh, S., Ahmed, Z., Stevenson, M., Briggs, S.S., Oupicky, D., Barrett, L.B., Spice, R., Kendall, M., Berry, M., Preece, J.A., Logan, A. & Seymour, L.W. (2005). A Versatile Reducible Polycation-Based System for Efficient Delivery of a Broad Range of Nucleic Acids. *Nucleic Acid Research,* Vol.33, No.24, (May 2005), pp. e86, ISSN 0305-1048

Ryan, K.J. & Wente, S.R. (2000). The Nuclear Pore Complex: A Protein Machine Bridging the Nucleus and Cytoplasm. *Curruent Opinion in Cell Biology,* Vol.12, No.3, (June 2000), pp. 361-371, ISSN 0955-0674

Sato, T., Ishii, T. & Okahata, Y. (2002). In Vitro Gene Delivery Mediated by Chitosan. Effect of pH, Serum, and Molecular Mass of Chitosan on the Transfection Efficiency. *Biomaterials,* Vol.22, No.15, (August 2002), pp. 2075-2080, ISSN 0142-9612

Sirirat, C., Lobo, B.A., Koe, G.S., Koe, J.G. & Middaugh, C.R. (2003). Biophysical Characterization of PEI/DNA Complexes. *Journal of Pharmaceutical Science,* Vol.92, No.8, (August 2003), pp. 1710-1722, ISSN 0022-3549

Son, S., Singha, K. & Kim, W.J. (2010). Bioreducible BPEI-SS-PEG-cNGR Polymer as a Tumor Targeted Nonviral Gene Carrier. *Biomaterials,* Vol.31, No.24, (August 2010), pp. 6344-6354, ISSN 0142-9612

Soundara, M.D. & Oupicky, D. (2006). Polyplex Gene Delivery Modulated by Redox Potential Gradients. *Journal of Drug Target,* Vol.14, No.8, (September 2006), pp. 519-526, ISSN 1061-186X

Suh, J., Wirtz, D. & Hanes, J. (2003). Efficient Active Transport of Gene Nanocarriers to the Cell Nucleus. *Proceedings of the National Academy of Sciences of the United States of America,* Vol.100, No.7, (April 2003), pp. 3878-3882, ISSN 0027-8424

Sun, Y.X., Zeng, X., Meng, Q.F., Zhang, X.Z., Cheng, S.X. & Zhuo, R.X. (2008). The Influence of Rgd Addition on the Gene Transfer Characteristics of Disulfide-Containing Polyethyleneimine/DNA Complexes. *Biomaterials,* Vol.29, No.32, (November 2008), pp. 4356-4365, ISSN 0142-9612

Szoka, F.C.J. (1993). Polyamidoamine Cascade Polymers Mediate Efficient Transfection of Cells in Culture. *Bioconjugate Chemistry,* Vol.4, No.5, (September-October 1993), pp. 372-379, ISSN 1043-1802

Tang, M.X., Redemann, C.T. & Szoka, F.C.J. (1993). In Vitro Gene Delivery by Degraded Polyamidomaine Dendrimers. *Bioconjugate Chemistry*, Vol.7, No.6, (November-December 1993), pp. 703-714, ISSN 1043-1802

Thanou, M., Florea, B.I., Geldof, M., Junginger, H.E. & Borchard, G. (2002). Quaternized Chitosan Oligomers as Novel Gene Delivery Vectors in Epithelial Cell Lines. *Biomaterials*, Vol.23, No.1, (January 2002), pp. 153-159, ISSN 0142-9612

Thomas, C.E., Ehrhardt, A. & Kay, M.A. (2003). Progress and Problems with the Use of Viral Vectors for Gene Therapy. *Nature Review Genetics*, Vol.4, No.5, (May 2003), pp. 346-358, ISSN 1471-0056

Thomas, M. & Klibanov, A.M. (2002). Enhancing Polyethylenimine's Delivery of Plasmid DNA into Mammalian Cells. *Proceedings of the National Academy of Sciences of the United States of America*, Vol.99, No.23, (November 2002), pp. 14640-14645, ISSN 0027-8424

Tomalia, D.A., Nylor, A.M. & Goddard, W.A. (1990). Starburst Dendrimers: Molecular-Level Control of Size, Shape, Surface Chemistry, Topology, and Flexibility from Atoms to Macroscopic Matter. *Angewandte Chemie International Edition*, Vol.29, No.2, (February 1990), pp. 138-175, ISSN 1433-7851

Verma, I.M. & Somia, N. (1997). Gene Therapy-Promises, Problems and Prospects. *Nature*, Vol.389, No.6648, (September 1997), pp. 239-242, ISSN 0028-0836

Wang, J., Mao, H.-Q. & Leong, K.W. (2001). A Novel Biodegradable Gene Carrier Based on Polyphosphoester. *Journal of the American Chemical Society*, Vol.123, No.38, (September 2001), pp. 9480-9481, ISSN 0002-7863

Wiethoff, C.M. & Middaugh, C.R. (2003). Barriers to Nonviral Gene Delivery. *Journal of Pharmaceutical Science*, Vol.92, No.2, (February 2003), pp. 203-217, ISSN 0022-3549

Wu, G., Fang, Y.Z., Yang, S., Lupton, J.R. & Turner, N.D. (2004). Glutathione Metabolism and Its Implications for Health. *Journal of Nutrition*, Vol.134, No.3, (March 2004), pp. 489-492, ISSN 0022-3166

Wu, G.Y. & Wu, C.H. (1987). Receptor-Mediated in Vitro Gene Transformation by a Soluble DNA Carrier System. *Journal of Biological Chemistry*, Vol.262, No.10, (April 1987), pp. 4429-4432, ISSN 0021-9258

Xue, Y.N., Liu, M., Peng, L., Huang, S.W. & Zhuo, R.X. (2010). Improving Gene Delivery Efficiency of Bioreducible Poly(Amidoamine)S Via Grafting with Dendritic Poly(Amidoamine)S. *Macromolecular Bioscience*, Vol.10, No.4, (April 2010), pp. 404-414, ISSN 1616-5187

You, Y.Z., Manickam, D.S., Zhou, Q.H. & Oupicky, D. (2007). Reducible Poly(2-Dimethylaminoethyl Methaerylate): Synthesis, Cytotoxicity, and Gene Delivery Activity. *Journal of Controlled Release*, Vol.122, No.3, (October 2007), pp. 217-225, ISSN 0168-3659

Zhang, H.W. & Vinogradov, S.V. (2010). Short Biodegradable Polyamines for Gene Delivery and Transfection of Brain Capillary Endothelial Cells. *Journal of Controlled Release*, Vol.143, No.3, (May 2010), pp. 359-366, ISSN 0168-3659

Zugates, G.T., Anderson, D.G., Little, S.R., Lawhorn, I.E. & Langer, R. (2006). Synthesis of Poly(Beta-Amino Ester)S with Thiolreactive Side Chains for DNA Delivery. *Journal of the American Chemical Society*, Vol.128, No.39, (October 2006), pp. 12726-12734, ISSN 0002-7863

5

Stable Magnetic Isotopes as a New Trend in Biomedicine

Vitaly K. Koltover
Institute of Problems of Chemical Physics, Russian Academy of Sciences, Chernogolovka, Moscow Region, Russian Federation

1. Introduction

Diverse organisms possess the ability to perceive Earth's magnetic field, the strength of which is about 0.05 mT (Lohmann, 2010; Gould, 2010). There exist magnetotactic bacteria the ability of which to use geomagnetic fields for direction sensing is accomplished owing to the so-called magnetosomes, the specific nanometer-sized magnetite particles organized into chains within the cell (Komeili, 2007).

However, apart from external magnetic fields, another variety of natural magnetism is around, namely, magnetic fields of atomic nuclei of magnetic isotopes. Some of them produce intramolecular magnetic fields which are 10-100 times greater than terrestrial (Grant & Harris, 1996). This raises the question of whether living cells can perceive the difference between magnetic and non-magnetic isotopes of the same chemical element. There is also a practical issue of whether the cell can take advantage of the magnetic isotopes.

The present article is a mini-review of the works of our group in this direction. The premises for our research have been the findings of magnetic-isotope effect (MIE) in chemical and biochemical physics within recent years (Brocklenhurst, 2002; Buchachenko, 2009). Following the concept of "nuclear spin catalysis in biopolymer nanoreactors" (Koltover, 2007, 2008), in experiments with bacteria *Escherichia coli*, the commonly accepted microbial model, we have revealed that the cells enriched with magnetic ^{25}Mg demonstrate essentially higher viability by comparison to the cells enriched with the nonmagnetic isotopes of magnesium (Bogatyrenko et al., 2009a, 2009b; Koltover et al., 2012). Furthermore, in experiments with *Saccharomyces cerevisiae*, another standard cell model, we have revealed that the magnetic isotope of ^{25}Mg, by comparison to nonmagnetic isotope ^{24}Mg, is essentially more effective stimulator of the recovery processes in the yeast cells after short-wave UV irradiation. The rate of post-radiation recovery was found to be twice as good for the cells enriched with ^{25}Mg as compared to the cells enriched with nonmagnetic isotope (Grodzinsky et al, 2011). Thus, the magnetic-isotope effects have been revealed, for the first time, *in vivo*. It opens up a new way in biomedicine, based on the stable magnetic isotopes, namely, the novel preventive medicine including new, ^{25}Mg-based, anti-stress drugs as well as anti-aging and anti-radiation protectors.

2. *Ab initio*: Magnetic-isotope catalysis in chemistry and biochemistry

Apart from the energy control (the law of conservation of energy), any chemical reaction as electron-nuclear rearrangement of reactants into products is controlled by angular momentum, spin, of reactants. Namely, the total spin of reaction products must be identical to the total spin of reactants. This law of spin conservation immediately follows from quantum mechanics, from the fundamental and universal Pauli principle: no two electrons may occupy the same quantum state simultaneously (see, e.g., Brocklenhurst, 2002; Buchachenko, 2009).

Figure 1 illustrates how the law of conservation of spin gives control over reactivity of free radicals, R•. For example, a pair of free radicals, each with electron spin S=1/2, may form a chemical bond and the resultant diamagnetic molecule, the total electron spin of which S=0 (Fig. 1a). However, from the law of conservation of spin, it follows that the chemical bond between these two radicals may happen only if the spin state of the pair at collision is singlet, i.e., spins of two electrons are subtracted to give the net S=0 (spin multiplicity, 2S+1=1). If the spin state of the radical pair is triplet, i.e., if the electron spins are added up to give the net S=1 (spin multiplicity, 2S+1=3), then the radicals cannot react immediately.

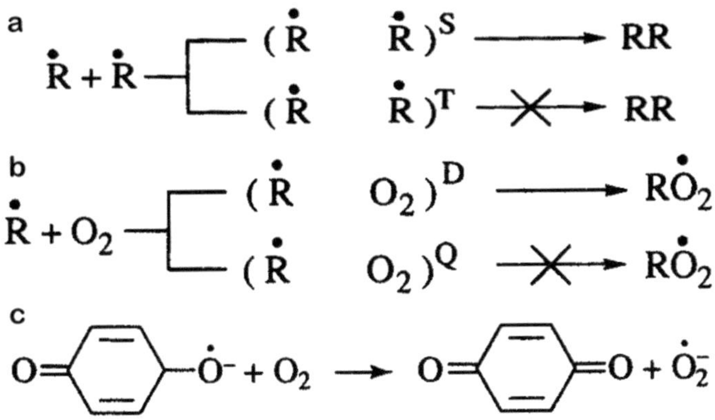

Fig. 1. Spin control in the chemical reactions of free-radical pairs.

In gas or liquid phase, with the time allotted for any collisions of radicals of order of 1 ns or less, neither spin-spin relaxation (in order of 10 ns) nor spin-lattice relaxation (in order of 100 ns) has time to fit the spin orientation. As a result, only one-quarter of encounters, with the radical pair in the singlet state, gives the recombination product while three-quarters of the initial radical pairs are inhibited from the reaction. Another example is presented on Fig. 1b. Namely, it is the reaction of free radical R• with oxygen, molecules of which are normally in the triplet spin state. The total spin, S, of this reagent pair can be 1/2 when the individual spins are subtracted (2S+1=1) or 3/2 when the individual spins are added up (2S+1=4). Meanwhile, for the product of the reaction $RO_2^{\bullet}$, peroxyl radical, S=1/2. Hence, from six possible spin states of the reactants only two states do not require a change in the total electron spin of the reactants and, therefore, are permitted for formation of $RO_2^{\bullet}$; other four states are forbidden for the reaction. A well-known reaction of mitochondrial ubisemiquinone with oxygen, in which $O_2^{\bullet-}$ is produced, exhibits a similar case (see, e.g., (Chance, 1979; Nohl et al., 1993). This reaction is permitted only from the doublet state of the reactants. Four quartet "channels" are forbidden by the law of the spin conservation (Fig. 1c).

To lift the ban on reactions forced by the law of spin conservation, spins of the reactants must be changed. Inasmuch as spin-orbital coupling is negligibly small in organic free radicals, magnetic fields are the only means which are able to change the spin states and, thereby, switch the reaction over spin-forbidden and spin-allowed channels. The probability of chemical reaction is a function of the parameters of magnetic interactions (Brocklenhurst, 2002; Buchachenko, 2009):

$$P=f(H;\ \omega;\ H_1;\ J;\ a;\ I;\ m_I;\ \mu_n)$$

In this equation H is external magnetic field (Zeeman interaction), ω and H_1 are frequency and amplitude of microwave magnetic fields. Correspondingly, acceleration of the free-radical reaction can be achieved through changes in the total electron spin of reactants by interaction with an applied external magnetic field. The parameter J is energy of the exchange interaction. Correspondingly, the reactions of organic free radicals or ion-radicals can be catalyzed via interaction of partners of the radical pair with a foreign, third spin carrier, like nitroxide radical. It is called "electron spin catalysis".

The above mentioned equation also contains parameters of hyperfine coupling a, nuclear spin I, nuclear spin projection m_I, and nuclear magnetic moment μ_n, i.e., the parameters of interactions of electron spins with magnetic nuclei which are known as the cause of the hyperfine splitting in EPR spectra of free radicals. Correspondingly, acceleration of the free-radical reactions can be achieved through changes in the total electron spin of reactants by interaction with magnetic fields of magnetic nuclei. This is known as "magnetic-isotope effect" (MIE): the reaction shows different reaction rates and different yields of products according to whether the reagents contain magnetic or nonmagnetic isotopes (Brocklenhurst, 2002; Buchachenko, 2009). While classical isotope mass effect selects isotopic nuclei in accordance with their masses, MIE selects isotopes by spin and magnetic moment. In action, MIE is a purely kinetic phenomenon and manifests itself as the dependence of the reaction rate on the nuclear spins of the reactants. Within recent years, MIE in chemistry has been discovered for a number of magnetic isotopes, among them H-D, ^{13}C, ^{17}O, ^{29}Si, ^{33}S, ^{73}Ge, $^{117,119}Sn$, $^{199,201}Hg$, and ^{235}U (Buchachenko, 2009). By analogy with "electron spin catalysis", the enhancement of the reaction rate by the nuclear spins of the reactants can be denoted as the "nuclear spin catalysis" (Koltover, 2007, 2008).

In biochemistry, MIE has been recently discovered for magnetic isotope of magnesium, ^{25}Mg, by A.L. Buchachenko and his group. It is generally known that energetic demands of every operation in living systems are met by molecules of ATP, be it eukaryotic cells of animals and plants or prokaryotic cells of bacteria. In aerobic organisms, most of ATP is produced in the so-called "oxidative phosphorylation". There are specific enzymes, "biomolecular nanoreactors", organized in the respiratory electron transport chains (ETC). Normal function of the ETC enzymes, be it mitochondrial nanoreactors in eukaryotic cells of animals or similar nanoreactors of bacteria cells, is in the transport of electrons, one by one, from the electron donor molecules to the end enzyme, cytochrome oxidase, from which the electrons are transferred to molecules of oxygen with two electron reduction of oxygen into water. Free energy released during the electron transport is used by the specific enzyme,

ATP-synthase, for synthesis of ATP from adenosine 5'-diphosphate (ADP) and inorganic phosphate (Nelson & Cox, 2008).

In the experiments with mitochondria isolated from the rat hearts, it has been revealed that the rate of oxidative phosphorylation with magnetic ^{25}Mg was 2-3 times higher than that with nonmagnetic ^{24}Mg and ^{26}Mg while no difference was found between the nonmagnetic magnesium nuclei (Buchachenko et al., 2005). It was also revealed that activity of phosphocreatine kinase and phosphoglycerate kinase, for which ions of Mg^{2+} serve obligatory cofactors, was essentially higher with magnetic ^{25}Mg than with ^{24}Mg and ^{26}Mg. Again, no difference in efficiency between the nonmagnetic magnesium nuclei was found in these experiments. Thus, there have been the very first evidences of MIE in biochemical reactions *in vitro*. Furthermore, the same research group has discovered MIE of calcium. The activity of phosphocreatine kinase with Ca^{2+} ions of magnetic nuclei ^{43}Ca was found to be twice higher than with Ca^{2+} ions of nonmagnetic nuclei ^{40}Ca (Buchachenko et al., 2011).

Factual evidence of MIE, on its own, indicates that there is a spin-selective "bottle-neck" of the process under investigation. The hypothetic mechanism of the acceleration of oxidative phosphorylation by the nuclear spin of ^{25}Mg suggested by this group is as follows (Fig. 2). Namely, they suggested a reversible transfer of electron density in the active center of ATP-synthase from the terminal anion phosphate group of ADP to Mg^{2+}-cation. It produces a virtual ion-radical pair, Mg^{+}-adenosine phosphate radical in the singlet spin state.

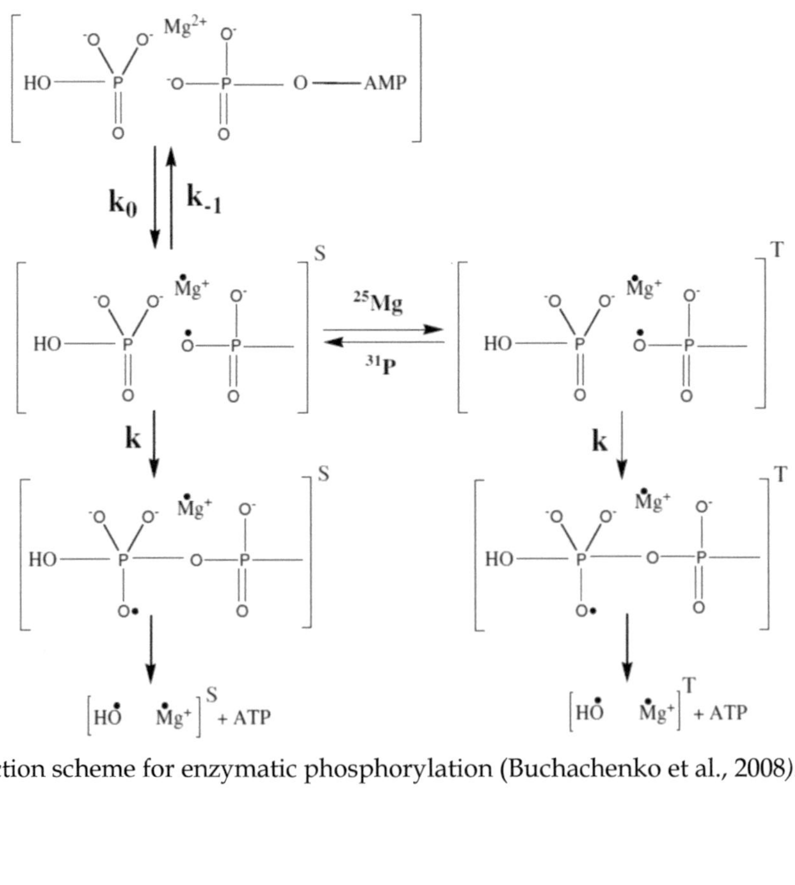

Fig. 2. Reaction scheme for enzymatic phosphorylation (Buchachenko et al., 2008).

Due to the hyperfine interaction of the unpaired electron with the nuclear spin of ^{25}Mg, the state of this virtual pair is converted from the short-lived singlet (total electron spin of the pair, S=0) into the long-lived triplet (S=1) in which the yield of the reaction of the ATP synthesis correspondingly increases (Buchachenko et al., 2008). A similar spin-selective ion-radical pair of Ca^+ with the phosphate radical of adenosine has been suggested to explain MIE of ^{43}Ca (Buchachenko, 2011).

It should be mentioned that the hypothesis about a possible key-role of such virtual ion-radical pairs in oxidative phosphorylation and energy transformation processes has long been stated (Blumenfeld & Koltover, 1972). Within the context of modern bioenergetics, which postulated a proton electrochemical gradient across the mitochondrial membrane as the energy-rich intermediate of oxidative phosphorylation in the "molecular motors" (see, e.g., Nelson & Cox, 2008), another plausible explanation for the MIE can be proposed. It is reasonably to suggest that the proton electrochemical gradient poses conformational pressure in the catalytic center of ATP-synthase, generating electronic-conformational excitation in the ADP-Mg^{2+} complex. This would essentially increase reactivity of the adenine base of ADP to phosphate (Koltover et al., 1971; Blumenfeld & Koltover, 1972). The nuclear spin of ^{25}Mg (or ^{43}Ca) can provoke the transition of the exited ADP-Mg^{2+} complex from the singlet state to the triplet state the lifespan of which is longer, thereby providing more time for the reaction of the ATP synthesis from ADP and P_i. Thus, the detailed mechanism of the magnetic-isotope catalysis in bioenergetics remains to be cleared.

3. *In situ*: Magnetic-isotope catalysis in living cells

There is a great variety of chemical elements in biomolecular nanoreactors of living cells (see Table 1). Certain of them are only represented by magnetic isotopes, among them - hydrogen, nitrogen, sodium, phosphorus, potassium, manganese and so on. However, there are chemical elements which have both kinds of stable isotopes, nonmagnetic and magnetic ones, among them - carbon, oxygen, magnesium, calcium, iron, zinc and others (Table 1). Correspondingly, these are the elements which are required to search for magnetic-isotope effects in living cells.

In this regard, magnesium is of particular interest. There are three stable isotopes of magnesium, ^{24}Mg, ^{25}Mg and ^{26}Mg with natural abundance about 79, 10 and 11 %. Among them, only ^{25}Mg has the nuclear spin (I=5/2) that produces the magnetic field. Two other isotopes are spinless (I=0) and, hence, produce no magnetic fields (Grant & Harris, 1996). As the most abundant intracellular divalent cation, Mg^{2+} is essential to regulate numerous cellular functions and enzymes. Ions of Mg^{2+} serve as obligatory cofactors in catalytic centers of many enzymes including ATP-synthase as the primary producer of ATP in mitochondria, chloroplasts, bacteria and archaea (Nelson & Cox, 2008). Moreover, a novel role for Mg^{2+} as an intracellular second messenger has been recently discovered (Li et al., 2011). Besides, the difference in masses between the isotopes of magnesium is much less, in percentage term, than that for the isotopes of carbon, for example, thereby minimizing the classical mass-isotope effect.

Stable isotopes of magnesium, namely nutrient solutions highly enriched with ^{25}Mg or ^{26}Mg, have been used for many years as *in vivo* tracers to determine magnesium absorption in human subjects, animals and plants models (see., e.g., Coudray et al., 2006; Weatherall et al., 2006). It is reasonable that the problem of possible beneficial effects of the magnetic isotope

Nucleus	Natural abundance, in %	Nuclear spin (*I*), in units of $h/2\pi$	Magnetic moment (μ), in nuclear magnetons ($eh/4\pi Mc$)	Biological functions
^{1}H ^{2}H	99.984 0.016	1/2 1	2.79270 0.85738	Structure unit of water, biomolecules, mitochondrial bioenergetics, etc.
^{12}C ^{13}C	98.89 1.11	0 1/2	0 0.70216	Structure unit of biomolecules, intracellular cation (HCO_3^-), extracellular buffer (HCO_3^-/H_2CO_3)
^{14}N ^{15}N	99.635 0.365	1 1/2	0.40357 –0.28304	Structure unit of amino acids, nucleic acid, etc.
^{16}O ^{17}O ^{18}O	99.759 0.037 0.204	0 5/2 0	0 –0.18930 0	Structure unit of water and biomolecules, biological oxidation, etc.
^{19}F	100	1/2	2.6273	Structure unit of dental enamel
^{23}Na	100%	3/2	2.2161	Main extracellular cation, functional unit of transmembrane electrochemical potential, deposit component of bond tissue, etc.
^{24}Mg ^{25}Mg ^{26}Mg	78.7 10.13 11.17	0 5/2 0	–0.85471	Main intracellular cation, structure and functional unit of chlorophylls in photosynthesis, obligatory cofactor of Mg^{2+}-dependent enzymes, including oxidative phosphorylation, glycolysis, synthesis of DNA and RNA, etc.
^{28}Si ^{29}Si	92.21 4.7	0 1/2	–0.55477	Main structure unit of exoskeleton in radiolarian and diatomic algae
^{31}P	100	1/2	1.1305	ADP, ATP, nucleic acids, phospholipids, P_i as main anion, P_i as regulator factor of transcription and translation, etc.
^{32}S ^{33}S ^{34}S	95.02 0.74 4.22	0 3/2 0	0.64274	Main intracellular anion (SO_4^{2-}), structure unit of cystine, cysteine and methionine, and glutathione, etc.
^{35}Cl ^{37}Cl	75.4 24.6	3/2 3/2	0.82089 0.68329	Main intracellular anion
^{39}K ^{41}K	93.08 6.91	3/2 3/2	0.39094 0.21453	Main intracellular cation, functional unit of transmembrane electrochemical potential, etc.
^{40}Ca ^{42}Ca ^{43}Ca ^{44}Ca ^{48}Ca	96.97 0.64 0.13 2.06 0.18	0 0 7/2 0 0	–1.3153	Main structure unit of bond tissue ($Ca_3(PO_4)_2$), regulator of membrane ion channels, Ca^{2+}-dependent myosin and other ATPases, etc.

Nucleus	Natural abundance, in %	Nuclear spin (I), in units of $h/2\pi$	Magnetic moment (μ), in nuclear magnetons ($eh/4\pi Mc$)	Biological functions
^{55}Mn	100	5/2	3.4610	Microbial Mn-superoxide dismutase, glutamine synthase, rat liver pyruvate carboxylase, etc.
^{54}Fe ^{56}Fe ^{57}Fe ^{58}Fe	5.82 91.66 2.245 0.33	0 0 1/2 0	<0.05	Heme- and non-heme proteins of electron transport, Fe-superoxide dismutase, etc.
^{59}Co	100	7/2	4.6388	Vitamin B_{12}, biosynthesis of heme proteins, etc.
^{63}Cu ^{65}Cu	69.09 30.91	3/2 3/2	2.2206 2.3790	Cytochrome oxidase, Cu,Zn-superoxide dismutase, ceruloplasmin, laccase, ascorbate oxidase, monoamine oxidase, etc.
^{64}Zn ^{66}Zn ^{67}Zn ^{68}Zn	48.6 27.9 4.12 18.8	0 0 5/2 0	0.8735	Cu, Zn-superoxide dismutase, DNA-polymerase, carbonic anhydrase, alcohol dehydrogenase, pyruvate dehydrogenase, pyruvate carboxylase, aldolase, etc.
^{75}As	100	3/2	1.4349	Activator of glycerylaldehyde phosphate dehydrogenase
^{74}Se ^{76}Se ^{77}Se ^{78}Se ^{80}Se ^{81}Se	0.87 9.02 7.58 23.52 49.82 9.19	0 0 1/2 0 0 0	0.5333	Structure-functional unit of glutathione peroxidase
^{92}Mo ^{94}Mo ^{95}Mo ^{96}Mo ^{97}Mo ^{98}Mo ^{100}Mo	14.84 9.25 15.92 16.68 9.55 24.13 9.63	0 0 5/2 0 5/2 0 0	 –0.9099 –0.9290 	Structure-functional unit of flavoproteins, nitrogenase, nitrate reductase, sulphite oxidase, xantine oxidase, etc.
^{127}I	100	5/2	2.7939	Structure-functional unit of thyroid hormones

Table 1. Stable isotopes in biological systems.

was not posed in the cited papers. There have been attempts to use the non-radioactive isotopes to make the oxidative biomolecules more stable against free-radical oxidation. It was found that deuterated polyunsaturated fatty acids protect yeast cells against the toxic effects of lipid autoxidation products (Hill et al., 2011). Both natural isotopes of hydrogen,

^{1}H and ^{2}H (D), are magnetic ones, but they have the twofold difference in masses. Consequently, the observed protection should be ascribed to the mass-isotope effect instead of the magnetic isotopy. In the *in vitro* study of cleavage of deuterated DNA by the hydroxyl radical, the value of the kinetic effect was found to be close to 2, just the mass-ratio of hydrogen and deuterium. Similar mass-isotopic effects are, presumably, anticipated for stable isotopes of nitrogen ^{15}N versus ^{14}N and for carbon ^{13}C that is 8% heavier than ^{12}C (Hill et al., 2011).

Recently, the smart methods of labeling of nematodes *Caenorhabditis elegans*, which are commonly used in gerontology, with ^{13}C and ^{15}N have been developed by feeding the worms with heavy isotope–labeled *Escherichia coli* (Fredens et al., 2011; Larance et al., 2011). However, with regards to the ideas of using the stable isotopes and their mass-isotope effects to stabilize cells against free-radical oxidation and, thereby, improve the living conditions and even extend the lifespan, there are doubts if high amounts of ^{13}C and ^{15}N in the numerous proteins, DNA, RNA, and other molecules of living cells will have, eventually, beneficial effects rather than harmful ones.

In searching beneficial isotope effects, including possible distinctions between the effects of magnetic and nonmagnetic isotopes, the very first step should be preparation of the cells enriched with different isotopes of magnesium. With this aim, we used bacteria *E. coli*, the commonly accepted cell model, and the growth media of the identical chemical composition with one exception, that they were supplemented with different isotopes of magnesium, magnetic ^{25}Mg and nonmagnetic ^{24}Mg or ^{26}Mg, as $^{25}MgSO_4$, $^{24}MgSO_4$ or $^{26}MgSO_4$ (Bogatyrenko et al., 2009a, 2009b; Koltover et al., 2012). The oxides of magnesium, ^{24}MgO, ^{25}MgO and ^{26}MgO, with isotope enrichment 99.9, 98.8 and 97.7 atom percent, correspondingly, were purchased from RosAtom, Russia. $^{24}MgSO_4$, $^{25}MgSO_4$ and $^{26}MgSO_4$ were prepared from the relevant oxides by using a standard acidic treatment with analytically pure sulphuric acid.

The pioneering studies have been done in Institute of Problems of Chemical Physics, RAS. Bacteria *E. coli*, strain BB, were cultivated in accordance with the standard design on the artificial liquid minimal M9-media, composed from 8 g of glucose, 2 g of NH_4Cl, 12 g of Na_2HPO_4, 6 g of K_2HPO_4, 1 g of NaCl in 750 ml of distilled water and 250 ml of tap water as the source of microelements. The growth media were supplemented with the isotopes of magnesium so that the final concentration of $^{24}MgSO_4$ or $^{25}MgSO_4$ was 2.2 mM per liter of the media. Cells were grown aerobically at 37 °C with shaking, harvested at the late growth ("stationary" phase at OD_{600} of about 0.5), and viability of the cells was tested as their ability to form colonies (colony-forming units, CFU) on the solid nutrient BCP agar using standard Petri's dishes.

The experimental data are presented in Fig. 3. The striking effect of the magnetic isotope, ^{25}Mg, has been detected when tallying up the colony forming units. The standard nutrient agar contains all components necessary for normal growth of cells, including magnesium. Nevertheless, the amount of CFU formed by the bacteria, which were previously grown on magnetic ^{25}Mg, has turned out to be about 40 percent higher in comparison with the bacteria, which were previously grown on nonmagnetic ^{24}Mg (Fig. 3a). Thus, the cells which have been previously enriched with the magnetic isotope of magnesium demonstrate the essentially higher viability in comparison to the cells enriched with the nonmagnetic isotope.

Another striking effect of the magnetic isotope of magnesium has been detected when measuring activity of superoxide dismutase (SOD), the main antioxidant enzyme of the cells. The cells enriched with ^{25}Mg demonstrate the reduced activity of SOD, about 40 percent, when compared to the cells enriched with ^{24}Mg (Fig. 3b). It is generally known that *E. coli* growing aerobically contain MnSOD and FeSOD (Imlay & Fridovich, 1991; Nelson & Cox, 2008). Inasmuch as the total SOD activity is normally adjusted to the intracellular level of $O_2^{\bullet -}$, the reduced level of SOD in the cells can be considered as evidence for lower production of $O_2^{\bullet -}$ as the failure by-product of cell respiration.

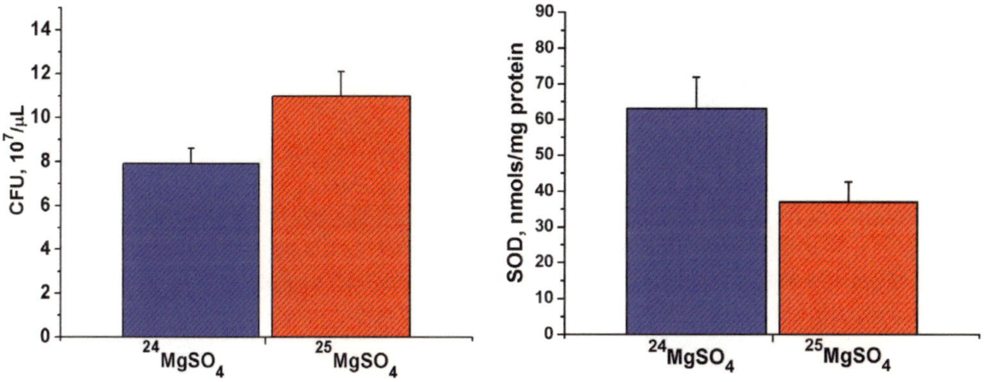

Fig. 3. The difference in effects of magnetic and nonmagnetic isotopes of magnesium on *E. coli*, strain BB (Bogatyrenko et al., 2009a). Left: ability of the cells enriched with magnetic ^{25}Mg or nonmagnetic ^{24}Mg to form colonies on the nutrition agar. Right: activity of superoxide dismutase in the cells enriched with magnetic ^{25}Mg or nonmagnetic ^{24}Mg. Data are indicated as $m \pm SD$, $N = 3$. The difference of the mean values (m) for magnetic ^{25}Mg *vs.* nonmagnetic ^{24}Mg is statistically significant at $P \leq 0.01$.

Thereafter, these experiments have been replicated in cooperation with the microbiologists of Orenburg State University using another strain of *E. coli*, K12TG1 (Koltover et al., 2012). After cultivation for 24 h in the artificial liquid minimal M9-medium without magnesium (and without tap water), the cells were suspended in the fresh M9-medium supplemented with different isotopes of magnesium as 0.26 g of $^{24}MgSO_4$, $^{25}MgSO_4$ or $^{26}MgSO_4$ per liter, and grown aerobically at 37 °C. To obtain reproducible results, three parallel experiments with each kind of the isotopes have been simultaneously performed, i.e. bacteria supplied with ^{24}Mg, or ^{25}Mg, or ^{26}Mg, three samples each, were simultaneously tested in the same experimental succession. After reaching the stationary phase with OD_{620} of about 0.5, the cells were tested for their ability to form colonies after inoculation on the agar plate's surfaces.

From the experimental results, shown on Fig. 4, one can see that the amount of CFU formed by the bacteria, previously grown on magnetic ^{25}Mg, has turned out to be almost twice higher in comparison with the bacteria which were previously grown on nonmagnetic ^{24}Mg and ^{26}Mg. Noteworthy, there has been no significant difference between nonmagnetic ^{24}Mg and ^{26}Mg in their effects on CFU. It gives evidence that there is the magnetic isotope effect of ^{25}Mg rather than a classical mass-dependent isotope effect of the magnesium isotopes.

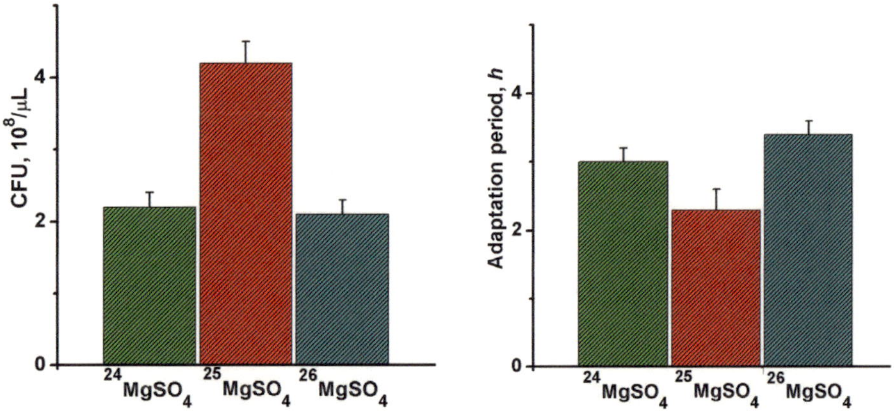

Fig. 4. The difference in effects of magnetic and nonmagnetic isotopes of magnesium on *E. coli*, strain K12TG1 (Koltover et al., 2012). Left: Ability of the cells enriched with magnetic ^{25}Mg or nonmagnetic ^{24}Mg and ^{26}Mg to form colonies on the nutrition agar. Data are $m \pm$ s.d., N =3. The differences of m for magnetic ^{25}Mg *vs.* nonmagnetic ^{24}Mg or ^{26}Mg are statistically significant at $P < 0.005$. Right: Length of adaptation period (lag-phase) on the liquid media supplemented with magnetic ^{25}Mg or nonmagnetic ^{24}Mg and ^{26}Mg. Data are $m \pm$ s.d., N =3. The differences of m for ^{25}Mg *vs.* ^{24}Mg or ^{26}Mg are statistically significant at $P \leq 0.02$.

In addition, kinetics of the cell growth was hourly monitored in the cited experiments, using optical density measurements at 620 nm with 96-cavity micro-plate reader "Uniplan" (Picon, Russia). The kinetic curves of cell biomass growth were typical for the bacterial growth involving slow adaptation period ("lag-phase") followed by exponential growth ("log-phase") when the cell mass quickly doubled and the stationary phase when the cell growth completed because of lack of the substrates (primarily, glucose). The striking observation for these experiments was that length of the lag-phase has turned out to be essentially shorter in the case when the cells were transferred on the media with magnetic isotope of ^{25}Mg in comparison with nonmagnetic ^{24}Mg and ^{26}Mg (Fig. 4b). Again, the nonmagnetic isotopes, ^{24}Mg and ^{26}Mg, were not differentiated by their effects on the cell growth. It means that adaptation of cells transferred from the pre-incubation media is limited by some enzymatic processes, efficiency of which is increased by magnetic nuclei of ^{25}Mg.

As it was cited above, magnetic isotope of ^{25}Mg more effectively performs the Mg^{2+} cofactor function for oxidative phosphorylation in the isolated mitochondria in comparison to ^{24}Mg or ^{26}Mg (Buchachenko et al., 2005). Energetic demands of every operation in prokaryotic cells of bacteria, as well as in eukaryotic cells of animals, are met by molecules of ATP. Hence, ATP as the main source of energy in living cells is most likely to be the limiting substrate for adaptation metabolism in the lag-phase. However, for the exponential phase of the cell growth, we have not found any significant differences in the rates of cell growth between magnetic ^{25}Mg and nonmagnetic ^{24}Mg and ^{26}Mg. The time of doubling of the cell mass was approximately the same, regardless of the type of magnesium isotopes (Koltover et al., 2012). This obviously suggests a different "bottle-neck" of metabolism in the exponential phase of growth with other, than ATP, limiting substrate and another limiting reaction independent upon the nuclear spin of magnesium.

Of special interest are searches for magnetic-isotope effects in the processes of recovery of cells from radiation injuries. The reason is that any factor, capable to influence on efficiency and reliability of cell nanoreactors, shows up itself most vividly under drastic conditions of post-radiation recovery (Koltover et al., 1980; Grodzinsky et al., 1987; Koltover, 1997).

We undertook the investigation of effects of magnetic and nonmagnetic isotopes of magnesium on post-radiation recovery of *S. cerevisiae* (Grodzinsky et al., 2011). The yeast cells (diploid strain MATα ade2Δ248 leu2-3,112 ura3-160,188 trp1 Δ:kanr) were cultivated on the standard nutrient liquid media M3 supplemented with $^{24}MgSO_4$ or $^{25}MgSO_4$. After three days of the cultivation under aerobic conditions at 30 ^{0}C, the cells were washed from the nutrient liquid and suspended in nutrient-free ("fasting") media, i.e., sterile phosphate buffer, pH 7.0. Then, the cells were irradiated by the short-wave ultraviolet light ($\lambda \approx 240$ nm, the dose ≈190 J/m^2), whereupon they were left in the nutrient-free water at 30 ^{0}C (with shaking) to study kinetics of the post-radiation recovery of the cells. For this kinetics study, the aliquots of the cells were periodically seeded on the standard nutrient (Petri dishes) and the cell survival was monitored by the standard CFU technique.

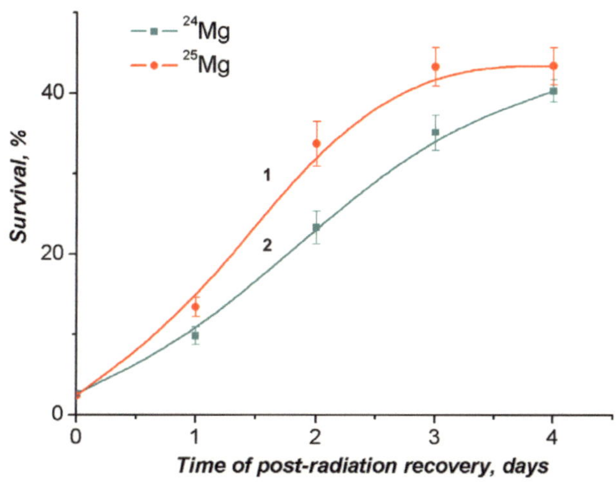

Fig. 5. The difference in effects of magnetic and nonmagnetic isotopes of magnesium on post-radiation recovery of *S. cerevisiae*, diploid yeast cells. The cells enriched with magnetic ^{25}Mg or nonmagnetic ^{24}Mg were irradiated by the short-wave UV-C light (λ = 240-260). Survival of cells was estimated as their ability to form colonies on the nutrition agar: 1 - recovery of the cells enriched with ^{25}Mg; 2 – recovery of the cells enriched with ^{24}Mg (Grodzinsky et al., 2011).

The survival of cells transferred to agar immediately after irradiation was not more than a few percent (Fig. 5). In this case the injured genetic structures in most of the cells could not be repaired before the onset of mitosis, and nonviable daughter cells were produced. Incubation in nutrient-free water, in which the cells do not divide, provides sufficient time for repair processes and leads to a corresponding increase in survival. From the kinetics curves represented on this figure we notice that the cells enriched with magnetic isotope of magnesium, ^{25}Mg, are recovered essentially more effectively than the cell enriched with the nonmagnetic ^{24}Mg.

It has been known that kinetics of recovery of yeast cells from radiation injuries may be represented by a function representing the reduction of the effective radiation dose, D_{eff}, with time:

$$D_{eff}(t)=D_o[k+(1-k)exp(-\beta t)]$$

In this model of A. Novick and L. Szilard, D_0 is the radiation dose, t is time of post-radiation recovery in the nutrient-free water, β is the recovery rate constant, and κ is the fraction of irreversible injuries (Grodzinsky et al., 1987; Koltover et al., 1980; Novick & Szilard, 1949).

Table 2 represents values of the kinetics parameters resulting from these experiments. While the fraction of irreparable injuries remained almost the same, the value of the rate constant β of the post-radiation recovery was twice higher for the cells enriched with ^{25}Mg than for the cells enriched with ^{24}Mg. This is decisive evidence that the magnetic isotope of magnesium essentially more effectively promotes the recovery of cells from radiation damages. Thus, we have, for the first time, documented the magnetic-isotope effect in radiation biology (Grodzinsky et al., 2011).

	β, h^{-1}	k
^{24}Mg	0.032 ± 0.003	0.70 ± 0.14
^{25}Mg	0.058 ± 0.004*	0.61 ± 0.12

Table 2. Effect of magnetic ^{25}Mg isotope on postradiation recovery of *S. cerevisiae*, diploid yeast cells, after short wave UV irradiation. *Difference between the means is significant at $P = 0.02$ (Grodzinsky et al., 2011).

One might suggest that the observed effects in our experiments with bacteria and yeast cells were caused by different levels of impurities in the growth media complemented with different isotopes of magnesium. However, it could hardly be the case. First, according to the mass-spectrometry data, amounts of contaminant elements in the stock solutions of the isotopes did not exceed 20-30 ppm, be it sulphate of ^{24}Mg, ^{25}Mg, or ^{26}Mg. Second, amounts of the contaminants that were administered in the liquid growth media with glucose and other basic components have significantly exceeded amounts of the same contaminants administered with much less additions of the isotope stock solutions. Besides, the impurities that were administered into the growth media from the basic components, as well as the element contents of the solid nutrient agar media, were obviously the same in all experiments, independently of the magnesium isotopes. Hence, one can disregard impurities as a possible reason of higher efficiency of magnetic ^{25}Mg than that of nonmagnetic ^{24}Mg and ^{26}Mg. It is apparent that the cells in the above cited experiments perceive the difference between magnetic and non-magnetic isotopes of magnesium, i.e., they perceive the nuclear spin's magnetic field of ^{25}Mg.

4. ***Future prospect*: Nuclear spin as factor of reliability in cell nanoreactors**

Thus, our data have documented, for the first time, the magnetic-isotope effect of magnesium-25 *in vivo*. Factual evidence of MIE, on its own, indicates that there is a spin-selective "bottle-neck" of the process under investigation. Ions of Mg^{2+} perform not only the cofactor functions in synthesis and hydrolysis of ATP. In addition, they have the impact on the structure-functional properties of RNA, RNA-polymerase, ribonuclease, and so on. Besides, there are the specialized proteins which regulate homeostasis and transport of Mg^{2+} in living cells (Romani, 2011). Moreover, ions of Mg^{2+} may work as the intracellular second messengers (Li et al., 2011). Up to date, however, there have not been findings of MIE except for the enzyme synthesis of ATP in the above cited papers (Buchachenko, et al., 2005, 2008, 2011). The similar MIE of ^{25}Mg is assumed to be in our experiments. Indeed, adaptation of cells to novel growth conditions requires a large variety of stress proteins to be synthesized and ATP, as the main source of energy in microbial cells, is most likely to be the limiting substrate for the adaptation metabolic processes. Similarly, a large variety of biosynthesis is required for recovery of cells from radiation injuries. It is reasonable to suggest that the kinetics of post-radiation recovery is also limited by spin-selective synthesis of ATP as the "bottle-neck". The post-radiation recovery proceeds with higher rate when the cell nanoreactors run on the magnetic isotope of magnesium, because the nuclear spin of ^{25}Mg catalyzes the ATP synthesis, hereby supplying the cells with more amount of ATP.

The lower level of superoxide dismutase (SOD) activity in the *E. coli* cells enriched with ^{25}Mg, by about 40 per cent when compared to the cells that were grown on nonmagnetic ^{24}Mg (Bogatyrenko et al., 2009a,b), can be also flow from the beneficial effect of the magnetic isotope on the ATP synthesis. The cell nanoreactors of oxidative phosphorylation have very ancient evolutionary origin and, hence, seem to be ones of the most reliable biomolecular machines. But yet their reliability ("robustness") characteristics are not perfect because these molecular machines experience conformational fluctuations (Grodzinsky et al., 1987; Koltover, 1997). It is well known that normal elementary acts of electron transfer on the electron transport chains, be it mitochondria or prokaryotes, alternate with random malfunctions when an electron, rather than waits for transport to the next enzyme of the electron-transport chain, goes directly to an adjacent oxygen molecule. Such an electron leakage results in production of $O_2^{\bullet-}$. Chemical products of $O_2^{\bullet-}$, the so-called reactive oxygen species (ROS), are toxic and initiate free-radical damages in the biopolymer nanoreactors (see, e.g., Chance et al., 1979; Koltover, 2009, 2010a). SOD catalyzes the reaction of dismutation of $O_2^{\bullet-}$ into hydrogen peroxide (H_2O_2) and oxygen, thus protecting cell structures from $O_2^{\bullet-}$ and its toxic chemical products. It makes its defense "job" in cooperation with two other specific enzymes, catalase and glutathione peroxidase, which catalyze decomposition of H_2O_2 into nontoxic reagents, namely H_2O and O_2 (see, e.g., Nelson & Cox, 2008, and references therein).

As a rule, the level of the SOD activity is adjusted to the intracellular level of $O_2^{\bullet-}$. If SOD activity decreases or increases, it normally reflects the relevant decrease or increase in the production of $O_2^{\bullet-}$ as faulty by-products of the electron-transport nanoreactors of oxidative phosphorylation (Imlay & Fridovich, 1991; Koltover, 2010a, 2010b, 2011). Hence, the lower level of SOD activity testifies the lower production of $O_2^{\bullet-}$ in the case when the cells are supplied with the magnetic isotope. As cited above, oxidative phosphorylation of ADP proceeds faster with ^{25}Mg by comparison with ^{24}Mg (Buchachenko et al., 2005). Since ^{25}Mg is more effective cofactor of oxidative phosphorylation, it transpires that the ATP synthase operates faster with the magnetic magnesium nucleus by comparison with the nonmagnetic ones. Meanwhile, under normal coupled conditions, the electron transport is subjected to the "backpressure" of the respiration-generated transmembrane electrochemical H^+-gradient. The partial dissipating of this gradient via the acceleration of ADP phosphorylation has been shown to result in the decreased production of $O_2^{\bullet-}$ (see, e.g., Mailloux et al., 2011). Hence, the higher rate of oxidative phosphorylation with magnetic ^{25}Mg, in comparison with nonmagnetic ^{24}Mg, through the reduction of the H^+-gradient "backpressure" should decrease the false electron leakage onto oxygen, thereby reducing the yield of free radicals $O_2^{\bullet-}$ as by-products of the electron transport.

From the point of view of chemical kinetics, with decrease in the rate of oxidative phosphorylation, there is the retardation of electron transport in the sites of the electron-transport chains which are coupled with phosphorylation of ADP. Inasmuch as the input of electron-transport nanoreactors becomes overflowed with electrons ("electron-transport jam"), the probability of electron leakage on oxygen increases. The more is acceleration of the ATP synthesis, the less is probability of the "jam". Hence, the yield of $O_2^{\bullet-}$ as the by-products of electron transport is bound to be much lower with ^{25}Mg-ADP by comparison with ^{24}Mg-ADP or ^{26}Mg-ADP.

Thus, with magnetic ^{25}Mg, the biopolymer nanoreactors of oxidative phosphorylation operate not only more effective but more reliable too, in comparison with their operation on non-magnetic isotopes ^{24}Mg and ^{26}Mg. Downgrading production of $O_2^{\bullet-}$, the magnetic isotope of magnesium produces, actually, a beneficial preventive antioxidant effect. This antioxidant effect of the nuclear spin, that ^{25}Mg favors the less production of reactive oxygen species, should obviously increase longevity of the electron-transport nanoreactors. Therefore, it can reveal itself in nature as the kinetic nuclear-spin selection of the favorable isotope, namely, the kinetic isotope enrichment with magnetic ^{25}Mg in the processes of recycling and regeneration of the electron-transport nanoreactors. For example, one can expect for the enrichment with the favorable magnetic ^{25}Mg in recycling and regeneration of cell mitochondria with aging of the cells (Koltover, 2007, 2010b).

Contrary, in the case of photosynthetic nanoreactors, one can predict undesirable pro-oxidant effect of the magnetic nucleus of ^{25}Mg. Indeed, it is known that the function of the vast majority of chlorophyll molecules (Chl), as the derivatives of the magnesium-protoporphyrin complexes, is to absorb light energy and transfer it to the specific energy sinks, the so-called reaction centers of the photosynthetic nanoreactors (Nelson & Cox, 2008). While performing this energy-transfer function, the light-exited Chl molecules are in the singlet state (^{1}Chl*, electron spin S=0). However, there is probability of the radiationless relaxation into the triplet state (^{3}Chl, S=1) followed by formation of singlet oxygen, $^1O_2$, the molecules of which are substantially more reactive by comparison with usual triplet O_2 molecules and, thereafter, produce oxidative damages. As nuclear spin of ^{25}Mg can catalyze the conversion of ^{1}Chl* into the triplet ^{3}Chl, one can expect for the higher yield of $^1O_2$ and, thereafter, more photodynamic damages in the cells with the chlorophyll molecules containing ^{25}Mg instead of the spinless ^{24}Mg or ^{26}Mg. Correspondingly, it is beyond reason to hope for selection of the magnetic ^{25}Mg, in the case of algae or green plants. Besides, the functional disadvantage of ^{25}Mg should be followed by increased synthesis of carotenoids and other natural antioxidants. Indeed, measurements of magnesium isotopic composition of the chlorophylls extracted from cyanobacteria and similar analysis of the chlorophyll forms in the leaves of English Ivy (*Hedera helix L.*) have revealed the isotope distribution following usual classical mass-isotope effect with no evidence for depletion or enrichment of ^{25}Mg (Black et al., 2007).

Apart from magnesium, there are many other elements which have both kind of stable isotopes, nonmagnetic and magnetic ones, amongst them - carbon, oxygen, calcium, iron and zinc (see Table 1). In passing, nuclear spin of ^{17}O should lift the spin ban over the reaction of the mitochondrial ubisemiquinone radicals with oxygen (see Fig. 1c), thereby catalyzing formation of $O_2^{\bullet -}$. As a result, one can expect for more reactive oxygen radicals and more free-radical damages due to ^{17}O in comparison with nonmagnetic ^{16}O and ^{18}O. The pro-oxidant action of ^{17}O can reveal itself as the selective enrichment of free-radical peroxidation products with this unfavorable magnetic isotope as compared with ordinary metabolites.

5. Conclusions and outlook

Factual evidence of magnetic isotope effect, on its own, indicates that the "bottle-neck" of the process under investigation is a free-radical or ion-radical reaction. Within the scope of free radical research, MIE can serve as the unique indicator to elucidate if the reaction under study proceeds through a free-radical or ion-radical pair as the key operand of the reaction.

Up to date, however, there have been no efforts to detect magnetic-isotope effects for other elements, except magnesium, in biopolymer nanoreactors.

Our experimental data have documented, for the first time, the beneficial magnetic-isotope effects of ^{25}Mg *in vivo*. Although the detailed mechanisms of the ability of the living cell to perceive magnetic properties of the atomic nuclei require further investigations, the "nuclear spin catalysis", as such, always and unambiguously indicates that the reaction under study is a spin-selective process with participation of paramagnetic intermediates, such as free radical pair, ion-radical pair or triplet state that undergo the spin conversion. Along this line the general principles of spin chemistry, amongst them - control of biochemical reactivity in living cells by the selective modification with stable magnetic isotopes, hold considerable promise. In part, the preventive antioxidant effect of ^{25}Mg opens the ways toward the novel biomedicine of anti-stress anti-aging drugs enriched with the magnetic-isotopes. The discovery of the magnetic-isotope effect in radiation biology opens up the way to the development of novel radio-protectors, based on the magnetic isotopy. Furthermore, inasmuch as the electron and nuclear spin moments can be changed by external magnetic fields, it makes possible to exert control over efficiency and reliability of biomolecular nanoreactors with the help of relevant magnetic and electromagnetic fields.

6. Acknowledgments

Financial support from Russian Foundation for Basic Research, projects no. 10-03-01203a and 10-04-90408-Ukr_a, is greatly acknowledged. I am grateful to L.V. Avdeeva-Tumanova, T.N. Bogatyrenko, & E.A. Kudryashova (Institute of Problems of Chemical Physics, RAS, Chernogolovka, Moscow Region), V.L. Berdinsky, D.G. Deryabin, E.A. Royba, & U.G. Shevchenko (Orenburg State University, Orenburg), T.A. Evstykhina & V.G. Korolev (Saint-Petersburg Institute of Nuclear Physics, RAS, Gatchina, Leningrad Region), and D.M. Grodzinsky & Y.A. Kutlakhmedov (Institute of Cell Biology and Genetic Engineering, Ukraine Academy of Sciences, Kyiv, Ukraine) for their fruitful collaboration. I am especially appreciated to Kathryn S. Tarasevich-Laukhina for her excellent assistance at the very beginning of this work.

7. References

Black, J.R., Yin, Q., Rustad, J.R., & Casey, W.H. (2007). Magnesium-isotopic equilibrium in chlorophylls. *J. Am. Chem. Soc.*, Vol. 129, 8690-8691.

Blumenfeld, L.A., & Koltover, V.K. (1972). Energy transformation and conformational transitions in mitochondrial membranes as relaxation processes. *Mol. Biol. (Moscow)*, Vol. 6, 161-166.

Bogatyrenko, T.N., Kudryashova, E.A., Tumanova, L.V., & Koltover, V.K. (2009a). Influence of different isotopes of magnesium on SOD activity level in the study of kinetics of growth of *E. coli*. *Proceedings of the V International congress on Low and Superlow Fields and Radiations in Biology and Medicine*, Saint-Petersburg, 29.06.2009-03.07.2009, p. 92.

Bogatyrenko, T.N., Kudryashova, E.A., Shevchenko, U.G., Tumanova, L.V., & Koltover, V.K. (2009b). Stable isotope of magnesium-25 as micronutrient Antioxidant. *Proceedings of the 3rd international conference on Nutrition, Oxygen Biology and Medicine*, Paris, 8-10 April, 2009, p. 57.

Brocklenhurst, B. (2002) Magnetic fields and radical reactions: recent developments and their role in nature. *Chem. Soc. Rev.*, Vol. 31, 301–311.

Buchachenko, A.L. (2009). *Magnetic Isotope Effect in Chemistry and Biochemistry*, Nova Science Publishing, New York.

Buchachenko, A.L., Kouznetsov, D.A., Arkhangelsky, S.E., Orlova, M.A., & Markarian, A.A. (2005). Spin biochemistry: magnetic ^{24}Mg-^{25}Mg-^{26}Mg isotope effect in mitochondrial ADP phosphorylation. *Cell Biochem. Biophysics*, Vol. 43, 243–252.

Buchachenko, A. L., Kouznetsov, D.A., Arkhangelsky, S.E., Orlova, M.A., & Markaryan, A.A. (2005). Magnetic isotope effect of magnesium in phosphoglycerate kinase phosphorylation. *Proc. Natl. Acad. Sci. USA*, Vol. 102, 10793-10796.

Buchachenko, A.L., Kouznetsov, D.A., Breslavskaya, N. N., & Orlova, M.A. (2008). Magnesium isotope effects in enzymatic phosphorylation. *J. Phys. Chem. B*, Vol. 112, 2548-2556.

Buchachenko, A.L., Kouznetsov, D.A., Breslavskaya, N. N., Shchegoleva, L.N., & Arkhangelsky, S.E. (2011). Calcium induced ATP synthesis: Isotope effect, magnetic parameters and mechanism. *Chem. Phys. Lett.*, Vol. 505, 130–134.

Chance, B., Sies, H., & Boveris, A. (1979). Hydroperoxide metabolism in mammalian organs. *Physiol. Rev.*, Vol. 59, 527–605.

Coudray, Ch., Feillet-Coudray, Ch., Rambeau, M., Tressol, J.C., Gueux, E., Mazur, A., & Rayssiguier, Y. (2006). The effect of aging on intestinal absorption and status of calcium, magnesium, zinc, and copper in rats: A stable isotope study. *J. Trace Elem. Med. Biol.*, Vol. 20, 73–81.

Fredens, J., Engholm-Keller, K., Giessing, A., Pultz, D., Larsen, M. R., Højrup, P., Møller-Jensen, J., & Færgeman, N.J. (2011). Quantitative proteomics by amino acid labeling in *C. elegans*. *Nature Methods*, Vol. 8, 845–847.

Gould, J.L. (2010). Magnetoreception. *Current Biol.*, Vol. 20, R431-R435.

Grant, D.M. & Harris, R.K. (Eds.). (1996). *Encyclopedia of Nuclear Magnetic Resonance*, Wiley, Chichester.

Grodzinsky, D.M., Evstyukhina, T.A., Koltover, V.K., Korolev, V.G., & Kutlakhmedov, Y.A. (2011). Effect of the magnetic isotope of magnesium, ^{25}Mg, on post-radiation recovery of *Saccharomyces cerevisiae*. *Proceedings of National Academy of Sciences of Ukraine*, No. 12 153-156.

Grodzinsky, D.M., Vojtenko, V.P., Kutlakhmedov, Y.A., & Koltover, V.K. (1987). *Reliability and Aging of Biological Systems*, Naukova Dumka, Kiev. 172 P.-book in Russ.

Hill, S., Hirano, K., Shmanai, V.V., Marbois, B.N., Vidovic, D., Bekish, A.V., Kay, B., Tse, V., Fine, J., Clarke, C.F., & Shchepinov, M.S. (2011). Isotope-reinforced polyunsaturated fatty acids protect yeast cells from oxidative stress. *Free. Radical Biol. Med.*, Vol. 50, 130-138.

Imlay, J.A., Fridovich, I. (1991). Assay of metabolic superoxide production in *Escherichia coli*. *J. Biol. Chem.*, Vol. 266, 6957-6965.

Koltover, V.K. (1997). Reliability concept as a trend in biophysics of aging. *J. Theor. Biol.*, Vol. 184, 157-163.

Koltover, V.K. (2007). Antioxidant and prooxidant effects of magnetic isotopes in biomolecular nanoreactors. *Free Radic. Biol. Med.*, Vol. 43, S. 67-68.

Koltover, V.K. (2008). Nuclear spin catalysis in free-radical biology: antioxidant effects of magnesium–25. *Free Radic. Biol. Med.*, Vol. 45, S. 12.

Koltover, V.K. (2009). Bioantioxidants: The systems reliability standpoint. *Toxicology and Industrial Health*, Vol. 25, No. 4-5, 295-299.

Koltover, V.K. (2010a). Antioxidant biomedicine: from free radical chemistry to systems biology mechanisms. *Russian Chemical Bulletin*, Vol. 59, No. 1, 37-42.

Koltover, V.K. (2010b). Reliability in nanoengineering: Nuclear spin of magnesium-25 as reliability factor in molecular and biomolecular nanoreactors. *Nanotechnology 2010*, Vol. 3, pp. 475-477, ISBN: 978-1-4398-3415-2, Danville (CA, USA): Nano Science and Technology Institute.

Kol'tover, V.K. (2011). Reliability of electron-transport membranes and the role of oxygen radical anions in aging: stochastic modulation of the genetic program. *Biophysics*, Vol. 56, No. 1, 125–128.

Kol'tover, V.K., Kutlakhmedov, Y.A., & Afanaseva, E.L. (1980). Recovery of cells from radiation-induced damages in the presence of antioxidants and the reliability of biological systems. *Doklady Biophysics*, Vol. 254, 159-161.

Koltover, V.K., Reichman, L.M., Yasajtis, A.A., & Blumenfeld, L.A. (1971). A study of spin-probe solubility in mitochondrial membranes correlated with ATP-dependent conformation changes. *Biochim. et Biophys. Acta*, Vol. 234, 296-310.

Koltover, V.K., Shevchenko, U.G., Avdeeva, L.V., Royba, E.A., Berdinsky, V.L., & Kudryashova, E.A. (2012). Magnetic isotope effect of magnesium in the living cell. *Doklady Biochemistry and Biophysics*, Vol. 442, No. 1-2, 12-14.

Komeili, A. (2007). Molecular mechanisms of magnetosome formation. *Annual Review of Biochemistry*, Vol. 76, 351–366.

Larance, M., Bailly, A.P., Pourkarimi, E., Hay, R.T., Buchanan, G., Coulthurst, S., Xirodimas, D.P., Gartner, A., & Lamond, A.I. (2011). Stable-isotope labeling with amino acids in nematodes. *Nature Methods*, Vol. 8, 849–851.

Li, F.Y., Chaigne-Delalande, B., Kanellopoulou, C., Davis, J.C. H. F., Douek, M.D.C., Cohen, J.I., Uzel, G., Su, H.C., & Lenardo, M.J. (2011). Second messenger role for Mg^{2+} revealed by human T-cell immunodeficiency. *Nature*, Vol. 475, 471-476.

Lohmann, K.J. (2010). Magnetic-field perception. *Nature*, Vol. 464, 1140–1142.

Mailloux, R J., Harper, M.-E. (2011). Uncoupling proteins and the control of mitochondrial reactive oxygen species production. *Free Radic. Biol. Med.*, Vol. 51, 1106-1115.

Nelson, D.L. & Cox, M.M. (2008). *Lehninger Principles of Biochemistry*, W.H. Freeman, New York.

Nohl, H., Koltover, V., & Stolze, K. (1993). Ischemia/reperfusion impairs mitochondrial energy conservation and triggers $O_2^{\bullet -}$ release as a by-product of respiration. *Free Radical Research*, Vol. 3, No. 18, 127-137.

Novick, A., Szilard, L. (1949). Experiments on light-reactivation of ultra-violet inactivated bacteria. *Proc. Natl. Acad. Sci. USA*, Vol. 35, 591-600.

Romani, A.M.P. (2011). Cellular magnesium homeostasis. *Arch. Biochem. Biophys.*, Vol. 512, 1-23.

Weatherall, A., Proe, M.F., Craig, J., Cameron, A.D., & Midwood, A.J. (2006). Internal cycling of nitrogen, potassium and magnesium in young Sitka spruce. *Tree Physiol.*, Vol. 26, 673-680.

Part 3

Medical Device Performance

6

Optical Fiber Gratings in Perspective of Their Applications in Biomedicine

Vandana Mishra and Nahar Singh
Central Scientific Instruments Organisation, Chandigarh
Council of Scientific and Industrial Research (CSIR),
New Delhi
India

1. Introduction

Optical fiber is a flexible, transparent *waveguide* or "*light pipe*" to transmit light; the latter, in its various forms and facets has caught the attention of humanity since prehistoric times. The ancient civilization used it as fire signals to communicate and later as a therapeutic and preventive tool for better health. In the modern era, the idea that light can be used for communication combined with the phenomenon of total internal reflection, gave rise to the concept of medium for light transmission. As a consequence, by the end of 19th century glass rods as illuminators were realized. Optical fibers were the next step as they are basically glass rods stretched very thin to become long and flexible. Gradual technological advances from 1920s when use of fiber for light transmission was first proposed, to 1980s resulted in glass fibers as the most ideal communication medium for enormous amount of data with lowest possible attenuation. Their inherent properties such as small size (have standard thickness of ~0.250 mm that can be less than that of surgical suture), biocompatibility, non-toxicity, chemical inertness and remote monitoring capability, make them quite lucrative for usage in the biomedical area. These fibers thus have diverse applications ranging from illumination to imaging, from phototherapy to precise surgery, from monitoring complex biomechanical dynamics to wearable smart sensors. In fact, after their first practical application in flexible endoscopes reported by Basil Hirschowitz in 1957 for illuminating and imaging internal organs of human body, the optical fibers have come a long way as sensors for various physiological parameters as well. This book chapter describes a special type of fiber optic tool, called fiber grating, its unique features with reference to potential applications in the field of biomedicine not only as in-fiber devices but also as sensing elements. (Mishra, 2011)

2. History

Although glass fibers as endoscope were being used for medical applications since 1950s, they had very limited applications because of their very high power loss (~1000 dB/km) and non availability of a compatible light source. The solution for the second problem came with the first Laser fabrication in 1960 by Maiman that had the potential to be ideal light source for optical fibers. After this and a revolutionary prediction in 1966 by [1]C.K. Kao and George Hockham that a purer fiber with 10 or 20 decibels of light loss per kilometer is possible to produce, there was a spurt of fiber based research activities worldwide. In 1970, two important breakthroughs happened i.e. first fiber with loss less than 20 dB[2]/km was fabricated and room-temperature operation of semiconductor laser was demonstrated; the latter became an ideal source for optical communication. (Hecht, 1999) By 1985, optical fibers with lowest possible loss (~0.2 dB/km) were being produced routinely. In that decade, research reached its pinnacle with optical fiber as a communication medium had been standardized to perfection. To quote *Philip Russell "standard fiber had become a highly respected elder statesman with a wonderful history but nothing new to say"*. (Russell, 2003). Thus subsequent advancement required exploring newer avenues like in-fiber devices and fiber optic sensors. In-fiber devices are essential for easier interconnection between fiber as communication medium and transmitter and receiver parts to complete efficient telecommunication. Other non-telecommunication applications, though started as spin-off, had been emerging simultaneously.

Concurrently, after the demonstration of fiberscopes further development in the quality of fibers and compact light sources resulted in a new offshoot of fiber optics i.e. fiber based sensors and other devices which were able to extract information about various aspects of human physiology by analyzing the reflected laser light sent and received through fibers. This method has been used in laser Doppler analysis of different cells. Study of scattered light is used to detect blood velocity to determine if sufficient blood is reaching vital organs. It can also detect the oxygen content of the blood. Miniature sensors at the end of an optical fiber were devised to measure pressures in the arteries, bladder, urethra and rectum. Some chemical analysis was also possible utilizing the phenomenon of luminescence. (Katzir 1989, Mishra et al 2009).

The discovery of grating formation in optical fibers by Hill and Coworkers in 1978 is a good example of serendipity! While studying non-linear properties of germanium doped silica by passing intense Argon ion laser radiation, they found an unexpected reflection and concluded that it was because of formation of Bragg reflection gratings inside the fiber core. This formation was attributed to interference between forward propagating wave and back reflected radiation from the far end of the fiber resulting in standing wave pattern. A refractive index distribution with the same periodicity as the interference pattern is thus created in the fiber core. This periodic perturbation of refractive index is a result of [3]*'photosensitivity'* phenomenon in certain kind of doped fibers. Introducing a variation of refractive index with periodicity on the scale of wavelength of light alters the light-matter interaction like a grating and results in selective reflection of light. Initially this phenomenon was just a scientific curiosity but after its first practical demonstration by Meltz in 1989,

---

[1] A part of 2000 Nobel Prize in Physics was awarded to Z.I. Alferov for his invention of semiconductor laser and that of 2009 to C.K. Kao for his "groundbreaking achievements concerning the transmission of light in fibers for optical communication"

[2] The dB(decibels) is related to the ratio of output optical power from an optical fiber to the input optical power; If an input power P1 results in an output power P2, the loss in decibels is given by;

α dB = 10 x Log_{10} (P1/P2)

[3] photosensitivity: the change in optical properties of material on exposure to light

there had been explosion of research activity in this field. (Kashyap, 2009, Othonos & Kallis, 1999)

3. Fiber gratings

Fiber [4]gratings are one of the simple intrinsic fiber devices that can reflect, filter or disperse light passing through them, suitable not only for communication applications but also finding their foothold as a fascinating sensor element in diverse areas. Figure 1 shows the schematic representation of a typical single mode fiber with fiber grating inscribed in its core.

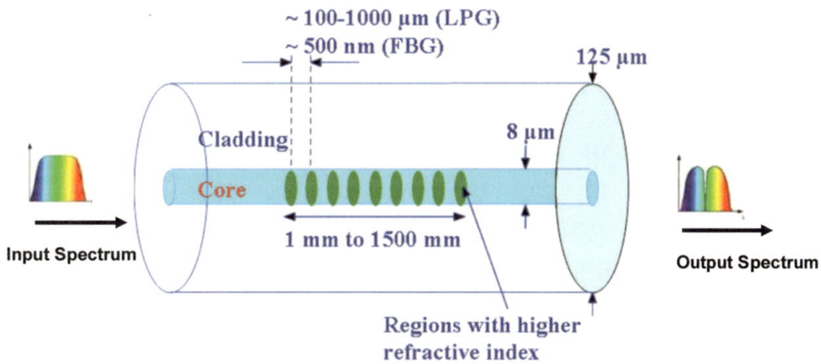

Fig. 1. Schematic representation of a fiber grating

3.1 Special features

Optical fibers are long and flexible waveguides generally made of fused silica (amorphous silicon dioxide, SiO_2) the most abundant mineral found in the earth crust. Though plastic optical fibers made of polymers are also quite practical and inexpensive for some applications, our focus here is on silica fibers which are the backbone of optical communication today. Their potential as efficient sensor elements was recognized simultaneously because of the unique advantages they offer, such as:

i. Small size and geometric versatility: with diameter in the range of 125 -500 micrometers, fiber sensors can be configured in arbitrary shapes and offer a great choice in space-restricted or hard to reach environments.
ii. Common technology base for multiparameter sensing: a single fiber can be used to sense various physical perturbations (acoustic, magnetic, temperature, strain, pressure, rotation, etc.)
iii. Compatibility with communication fiber facilitates telemetry
iv. Absence of crosstalk between close fibers suggests that different sensors can be housed in close vicinity e.g. a catheter.
v. A single electro-optic unit can be utilized for all the sensors, naturally with an appropriate illumination, detection and signal-processing scheme.

[4] "Grating" here is used in the sense of "diffraction grating" which is usually a periodic structure having very fine parallel grooves or slits and used to produce optical spectra by diffraction of light.

vi. Optical fibers have very low attenuation and thus long fiber links can be used to remotely monitor and control the sensing parts without disturbing the patient while keeping all the electronics away.
vii. Immunity to electromagnetic/ radiofrequency interference and chemically inert nature: due to their dielectric construction, they can be used in high voltage, electrically noisy, high magnetic field, high temperature, corrosive, or other harsh environments

Along with these, fiber gratings offer some added features unique to them like self-referencing as the information is wavelength encoded and ease of multiplexing, facilitating distributed sensing, that make them more valuable as sensors. Their potential as strain and temperature sensors array in various concrete, metal and composite structures has been established much earlier and they have been implemented successfully for *structural health monitoring (SHM)* in the arena of civil and aerospace, oil & gas exploration wells, power systems monitoring, etc. (Rao 1997, Majumdar 2008, Tiwari 2009)

3.2 Relevance in biomedicine

Because the fiber gratings are made of dielectric glass, they are inherently immune to electromagnetic interference (EMI) and can be safely used without any electrical or chemical obstruction in a clinical setup. Their miniature size, flexibility, and lightness allow easy insertion in catheters/ needles making possible localized measurements inside blood and tissues. Also, multiple sensors can be accommodated on a single fiber, working independently of the other. Silica is chemically inert and fulfills the biocompatibility criterion. (Davis 1972, Yang 2003) As fibers are intrinsically safe for the patient i.e. their use produces no immunity response from human defense system, they can be used for *in- vivo* measurements and can be left in their position for repeated or continuous monitoring.

Although all these qualities are quite exciting there are some issues coming into the way of fiber grating technology to become accepted in healthcare systems that remain to be dealt with. For example, howsoever small or non-intrusive these sensors are for the patient, their read out units need to be connected through a fiber link which is not so practical in the modern world of wireless technology. With the emergence of smaller and faster interrogators one possible way out can be the development of very small wearable interrogator system with no external fiber links. Another issue is information extraction process that needs to be standardized for each application. Fiber gratings provide information in terms of wavelength shift (given by equations 2-5 in the next section) that can be due to changes in various external parameters e.g. strain, temperature, pressure, refractive index (RI) etc. when the sensor is being used for one specific measurand (strain for example) it should be unaffected by all other parameters (temperature, RI etc.) to minimize error. This can be achieved by using optimized packaging/transducer. Also, if one sensor is being used for multiparameter sensing, it will require customized interrogation system and software tool to discriminate the effect of each parameter. Practical solutions of all these issues require multidisciplinary approach with a synergy between various experts e.g. doctors, physicists and engineers.

4. Working principle

Optical fiber consists of an inner dielectric core surrounded by cladding of another dielectric material of slightly lower refractive index than that of core. In the simplest case, the waveguide effect is often explained as resulting from repeated total internal reflection (TIR) of light rays at the core–cladding interface. The fiber can then guide all light impinging the input face under certain conditions. Standard fibers have core and cladding refractive indices constant along the fiber length, but if a periodic modulation of refractive index is created deliberately along the fiber core it results in formation of fiber gratings.

The periodic variation of refractive index in the fiber core is created mostly by its exposure to Ultraviolet (UV) radiation through a proper mask. These gratings are categorized as short period or fiber Bragg gratings (FBGs) and long period gratings (LPGs). In an FBG a narrow band of wavelength is reflected and there is a corresponding drop of intensity in the transmitted spectrum while LPGs work as wavelength dependant loss elements exhibiting multiple loss resonance bands in the transmitted spectrum. Though their ability to filter certain wavelengths was a major attraction in the field of telecommunication, their potential as sensing devices was recognized simultaneously. This is because the filtered wavelength is a function of its effective refractive index as well as the period of the grating and any variation in these parameters results in shift of wavelength which can be detected easily using an optical spectrum analyzer (OSA) or an interrogator. Apart from basic FBG and LPGs there are some other distinct grating types that are formed when deliberate nonuniformity in the refractive index profile is introduced. *Blazed* or *tilted* FBGs are resulted when grating planes are at an angle with the fiber axis and are used mostly for mode conversion. A tilted grating can be designed in a way that its' core mode is coupled with some of the cladding modes so that the Bragg Wavelengh becomes a function of ambient refractive index and thus can be used for refractive index sensing. Another type is *chirped* FBG in which the grating period is aperiodic having a monotonic increase/decrease or non-uniformity longitudinally. In this type of gratings each point has different Bragg wavelength and hence its spectrum can be used to monitor a parameter profile by distributed sensing (Kashyap, 2009).

In an FBG, the guided light is scattered by each interface of different refractive index regions in the core and for a wavelength which satisfies Bragg Condition, the scattered light adds up constructively resulting in back reflection with a central wavelength (λ_B) given by

$$\lambda_B = 2n\Lambda \text{ [Bragg Condition]} \tag{1}$$

Where Λ is the pitch or periodicity of the grating, n is the effective refractive index of the core and λ_B is the Bragg wavelength (Kashyap 2009, Othonos & Kallis 1999). Therefore, when light from a broadband source is launched in this FBG the spectral component defined by above equation is missing from the transmitted spectrum. Bragg wavelength is shifted if the effective refractive index, the grating periodicity or both are changed due to some perturbation; in fact both these parameters are directly influenced by strain and ambient temperature with the associated wavelength shift given as

$$\Delta\lambda_B = 2[\Lambda \frac{\partial n}{\partial l} + n \frac{\partial \Lambda}{\partial l}]\Delta l + 2[\Lambda \frac{\partial n}{\partial T} + n \frac{\partial \Lambda}{\partial T}]\Delta T \tag{2}$$

Where Δl is change in grating length due to strain and ΔT is change in ambient temperature. The first term on the RHS gives strain dependence while the second term indicates temperature dependence of the Bragg wavelength and an FBG sensor works by monitoring Bragg wavelength shift with one or both of these parameters.

LPGs couple fundamental guided core mode to different cladding modes. The loss resonance wavelength(s) at which this coupling takes place satisfies the phase matching condition i.e.

$$\lambda_i = [n_{eff}^{co} - n_{ieff}^{cl}]\Lambda \tag{3}$$

Where n_{eff}^{co} and n_{ieff}^{cl} are the effective refractive indices of the fundamental core mode and ith cladding mode respectively and Λ is the period of the LPG. Since effective index of a cladding mode is dependent upon the refractive index of the surrounding medium, any change in the latter alters the loss resonance wavelength(s). The influence of refractive index of the surrounding medium on the LPG wavelength(s) is expressed by the following equation,

$$\frac{d\lambda_i}{dn_{sur}} = \frac{d\lambda_i}{dn_{ieff}^{cl}} \cdot \frac{dn_{ieff}^{cl}}{dn_{sur}} \tag{4}$$

Where, n_{sur} is the refractive index of the surrounding medium.

Apart from the shift in loss resonance wavelengths, the variation in n_{sur} is also reflected as variation in intensity of the loss resonance peak, defined by the overlap integral *I* given by equation (5),

$$\mathrm{I} = \frac{\int_r \int_\phi \psi_{core} \psi_{clad}^{*} r dr d\phi}{\sqrt{\int_r \int_\phi \psi_{core} \psi_{core}^{*} r dr d\phi} \sqrt{\int_r \int_\phi \psi_{clad} \psi_{clad}^{*} r dr d\phi}} \tag{5}$$

Where, ψ s are the electromagnetic field components of the two coupling modes, r and Φ are radial and azimuthal co-ordinates respectively (Mishra et al 2005). Obviously, any change in cladding field distribution will affect the coupling strength. When the refractive index of the surrounding is varied, it will alter the cladding field distribution and hence the overlap integral. Since effective refractive index of a cladding mode and their coupling efficiency with the fundamental mode is dependent upon the refractive index of the surrounding medium, any change in the latter is easily detected using LPGs. This is the basic principle of an LPG based refractive index and concentration sensor. (James 2003, Patrick & Kersey 1998)

5. Fabrication technology

Direct inscription of submicron periodic pattern in optical fibers puts severe constraints in terms of stability and precision on the grating fabrication techniques. There are only a few methods, namely, the *interferometric*, the *phase mask*, and the *point-by-point* techniques that give consistently good quality gratings. (Othonos 1999)

In our laboratory, fiber gratings are produced by exposing core of a photosensitive fiber (StockerYale Inc.) to intense UV light from a KrF (Krypton Fluoride) excimer laser at 248nm wavelength. The primary acrylate coating is first removed from the fiber section that is to be exposed to UV light, using a thermo-mechanical stripper to avoid mechanical degradation. This uncoated fiber is further loaded onto the fiber mounting stage and an appropriate tension is applied to keep it straight, using load cell. After that the fiber is first made to approach close proximity of the mask within 0. 5mm and vertical & horizontal alignments are performed by continuously monitoring on a computer screen. To inscribe FBGs the light from the KrF laser is made incident on the fiber through a phase mask of a period as per the design of the FBG being inscribed and scans on the required length of the fiber to get the FBG inscription of the designed parameters e.g. (a typical value of 1060 nm periodicity phase mask is used in our lab to achieve the peak wavelength in the range of ~1550 nm). To achieve the FBG with more than 90% reflectivity, 7-8 scans were applied. For LPG fabrication point-by point method is used to expose fiber core to laser radiation. To monitor the FBG parameter within the appropriate design, optical spectrum analyzer with broadband ASE (**Amplified Stimulated Emission)** source is used while for LPGs white light source is employed. Figure 2 shows the experimental setup for grating fabrication.

Fig. 2. Grating fabrication set up

To protect the FBG from external environment and to provide mechanical strength, the stripped fiber section is immediately re-coated with acrylate while LPGs are left uncoated so that they remain sensitive to external refractive index. Care should be taken to keep the bare LPGs well protected; they should be either fixed in a working glass cell or kept properly covered when not in use. Finally, for stabilization of grating properties over long period of time, thermal annealing of inscribed gratings at high temperature (>150 °C) is carried out.

6. Current research scenario

The multiplexing, multi-parameter and minimally invasive sensing capabilities even in high electric/magnetic field environments of fiber grating sensors make them befitting for various applications in healthcare as evident by past and ongoing research activities all over the world. Some areas of current activities are described briefly in subsequent sub-sections.

6.1 Biomechanics

Biomechanics involves application of engineering mechanics to biological systems to better understand them and make use of conventional electrical strain gages (ESGs) as a tool. These ESGs, considered gold standard for strain measurement, consist of metallic parts and it is difficult to make them adhere on the surface of a bone or other biological tissues. Besides, working of an ESG depends upon measuring electric resistance that varies with applied strain and so it is not suitable in strong electric and magnetic field environment associated with medical appliances. Moreover, they cannot be made completely biocompatible. Fiber gratings have an edge over the conventional electrical gages even for *in-vivo* applications because they have smaller risk for infection and can also be used even on curved surfaces or in locations where use of a conventional gage is technically and medically not feasible.

A temperature independent array of FBGs with proper design can be used as pressure sensor to measure muscular strength of hands or weight profile of patients when used under foot. In a study started almost a decade earlier, FBG pressure sensors embedded in Carbon Fiber reinforced material were designed and investigated by a research group in Singapore for monitoring the foot pressure of diabetic patients (Hao et al 2003). These sensors give better results in terms of sensitivity, cost and compactness as compared to conventional foot pressure sensing systems. This concept along with development of multiple neural networks for continuous monitoring of various parameters simultaneously can be used for *human gait analysis*. Gait analysis is important for patients with cerebral palsy, neuromuscular disorders or diabetes and many hospitals worldwide now have gait labs both to design treatment plans, and for follow-up monitoring. Existing systems of gait analysis use large numbers of sensors or complex imaging systems, whereas several FBGs on a single strand of fiber with one interrogation unit can be embedded within materials forming a surface without loss of material strength. Recently, FBG sensors have been investigated for distributed tactile sensing (Cowie et al 2006, 2007) in a grid arrangement along with a neural network to detect the position and shape of a contacting load simultaneously in real- time.

Multiple FBG sensors in a single fiber can be bonded at strategic locations on the patients' bed to continuously monitor their movements from a remote station. The concept of a smart bed is under investigation in Singapore (Hao et al 2010). A series of 12 FBG sensors underneath the patient's mattress with suitable algorithm give pressure profile and respiratory rate of the patient while another set of gratings placed on top of the mattress detect heart rate. Existing methods require different techniques for each individual parameter while this single system shown schematically in figure 3, can monitor respiratory rate, heart rate, pressure points and occupancy of patient on bed in a continuous, non-intrusive and robust manner.

An ongoing project entitled, **"Intelligent adaptable surface with optical fiber sensing for pressure- tension relief" (IASIS)** of European Commission is incorporating FBG-based 2D and 3D sensing systems for utilization in "smart" fiber-based Human Machine Interfaces (HMI) employed in clinical beds, amputee sockets and wheelchair seating systems, targeting pressure ulcer and wound treatment. **(Pleros et al 2009).**

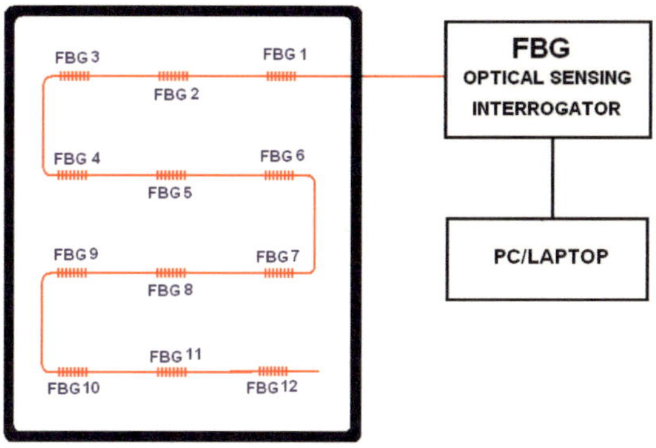

Fig. 3. Schematic of the FBG sensor array for smart bed

Recently some studies on the possible use of FBG as a strain gage in bones have been undertaken. Although in-vivo strain measurement in humans is not very common, the researchers at Hadassah University Hospital, Israel have reported use of instrumented bone staples made of electrical strain gages in some volunteers (Milogram 2000, 2003). Talaia et al (2007) have first reported use of FBG sensor array to study strains in fracture fixation of synthetic femur. As strain measurement on bone plates using ESG is technically difficult and not feasible, FBGs are good alternative to assess the stiffness of callus formation of fractured bones. Use of FBG sensors in place of ESGs have been validated to measure deformation in human cadaver femur bone specimen under in-vitro loading condition (Fresvig et al 2008). The effect of **decalcification** on strain response of a goat tibia was investigated *in-vitro* using FBG sensors by our group (Mishra et al 2010). In the investigation, two tibia bone samples were taken from the same animal. The FBG sensor was directly bonded onto the surface at the midpoint of the bone shaft, which is the most vulnerable point, using standard cynoacrylate adhesive and cured properly. One sample was decalcified in gradual steps while the other was kept in saline solution for a comparative study. Both the bone samples were strained by using three-point bending technique and corresponding Bragg wavelength shifts were recorded. Strain response of the decalcified and untreated bones was studied concurrently to monitor the effects of calcium loss and that of degradation with time. The strain generated for the same stress increased with greater degree of decalcification e.g. calcium loss of even 0.3906 gm (treatment 1) resulted in 1.3 times/ 24% more strain for the same load while calcium loss of 1 gm resulted in 50% increase in strain and when calcium loss was 2.78 gm the increase in strain reached more than 2000% i.e. 22 times more strain as compared to that before decalcification. Figures 4a &4b show the schematic of the experimental setup developed and the results obtained respectively. The Dexa results of bone samples matched with our results for mineral loss.

It is possible to measure strain less than 5 micro-strains accurately using this sensing technique, which can be indicative of the onset of decalcification. The objective of this investigation was to develop a different, efficient and safe method to estimate calcium levels in bones. The small size of fiber can be utilized to make strain staples much smaller then the existing ones so that it can be implanted using *minimally-invasive* surgical method, or, as this technology is still developing it may advance into *non-invasive* method in future.

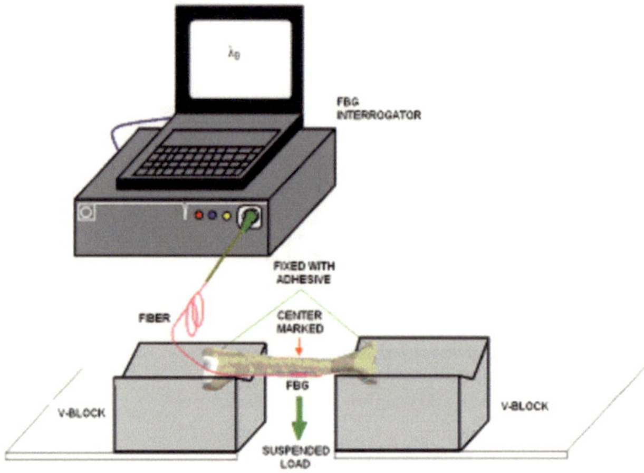

Fig. 4a. Schematic Illustration of the Experimental Set up

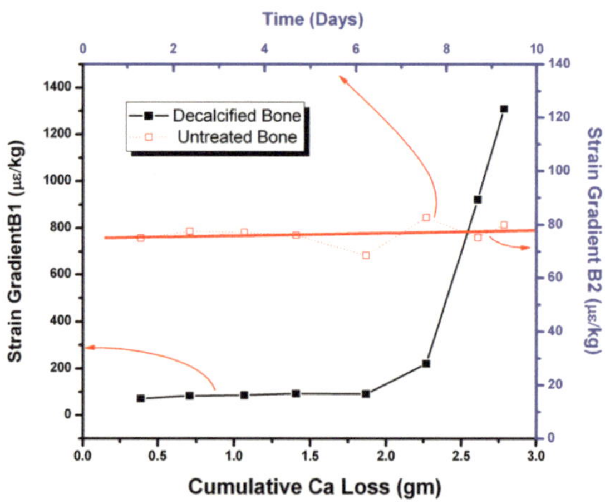

Fig. 4b. The Results of Bone Decalcification Study

First application of embedded FBG sensors in dental biomechanics was reported (Tjin et.al 2001) to monitor the force and temperature as a function of time in dental splints used by patients with obstructive sleep apnoea. This monitoring gives a clear indication of the compliance of the patient with regard to the proper usage of splint which is necessary for its effectiveness. In another type of investigation FBGs written in polarization maintaining fibers were used to monitor the drying of dental cement and the corresponding stress build-up (Ottevaere et al 2005). To measure strain at a mandible surface caused by impact loads on dental implants, an FBG sensor was employed by JCC Silva et al (2004) on a dried cadaveric mandible showing the feasibility of using FBG to monitor dynamic strain in complex biomechanical systems. A research group in Portugal has applied FBG sensors to assess the performance of dental implant system by measuring static and dynamic bone strains around it. Conventional techniques can not be used to measure strains inside bones (Schiller et al 2006). This study can lead to significant improvement in the design of dental implants.

Now-a-days, the use of custom-made mouthguards as preventive measures for persons participating in sports activities is being encouraged as they have an injury-preventing ability. To evaluate the performance characteristics of such mouthguards, no standard technique exists till date. A unique experimental scheme utilizing fiber Bragg gratings (FBGs) as distributed strain sensors is proposed and investigated by our group to estimate impact absorption capability of custom-made mouthguards. In the scheme, a pendulum based fixture with interchangeable impact load e.g. cricket, hockey and steel balls, was custom made for the investigation. Two sets of FBGs were used; one at the mouthguard surface and the other at similar position on a jaw model. The fixture was used to simulate impact using different balls released from varying angles and Bragg wavelength shifts of FBGs at mouthguard and that at the jaw model were recorded. Figures 5a & 5b show photograph of the jaw model, mouthguard with pendulum setup and change in FBG spectrum due to impact. It **is** obvious from the FBG response shown in figure 5b that for the same impact there is no detectable change in the Bragg wavelength of the FBG bonded on

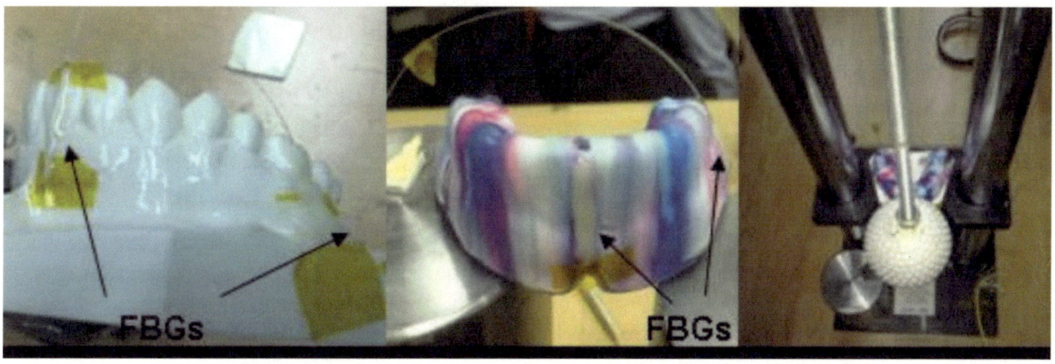

Fig. 5a. Photograph of the Experimental Set up for Mouthguard Experiment

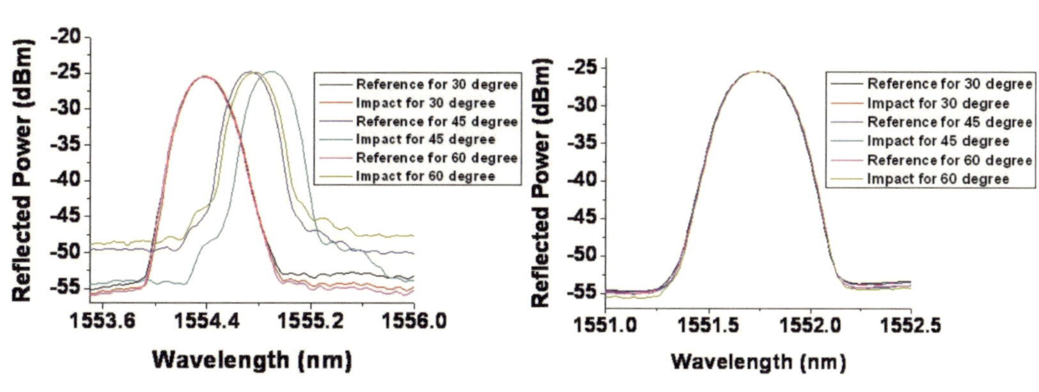

Fig. 5b. Spectral shift due to Ball Impact of FBG Sensors fixed on Mouthguard and on Jaw Model

the jaw model while FBG on the mouthguard has shown large wavelength shift with each impact. Relative Bragg wavelength shift with respect to each impact load determines the protection capability of the mouthguard. Due to multiplexing capability of FBGs, it is possible to fix multiple sensors in series at various points of mouthguard and denture to evaluate the effect of a single impact on different locations simultaneously. Impact tests on various locations can detect the vulnerable points where the mouthguard is less protective.

Through such studies it will be possible to quantify the level of protection and hence to predict the required modifications in the mouthguard. This research work thus, is important for the establishment of guidelines for design of safer mouthguards. (Tiwari et al 2011)

J Paul et al (2005) had suggested use of five FBGs written at different wavelengths to measure **handgrip strength** through a grip holder. Handgrip strength monitoring is rated as one of the top ten fitness tests to evaluate different physical and functional disorders related to healthcare. The conventional methods (dynamometer) are rough, uncomfortable and do not provide individual finger strengths; thus not suitable especially for rehabilitation programs.

Researchers at Nanyang Technological University, Singapore have reported an FBG based sensor in instrumented tibial spacer (ITS) to correct **misalignment during total knee replacement** surgery. The sensor, comprising of optical fibers with sampled chirped gratings inscribed on each fiber to generate a pressure profile, was embedded in a fiber-reinforced composite. During a total knee joint replacement procedure, the ITS sensor can slide in place of the prosthetic spacer. The femur can be rolled over the ITS sensor and the alignment checked from the pressure map displayed. Any misalignment can be corrected with repeated checking. After the measurements are taken and the required alignment achieved, the ITS sensor can be replaced by the actual tibial prosthetic spacer and the knee joint can be sutured (Mohanty et al 2007).

Methods were developed by Dennison et al 2007 to measure **intervertebral disc pressure** response to compressive load in five lumbar functional spine units, using FBG in a patented configuration. The pressure measurement with FBGs is less disruptive than the existing techniques. In an improved configuration FBG sensor placed in silicone filled needle were applied to intervertebral disc pressure measurements in a cadaveric porcine functional spinal unit and the results were in agreement to those obtained with the standard strain gage sensor. (Dennison et al 2009 1 &2).

Investigative study of FBG sensor for in-vitro **biomechanical properties of porcine tendons** was reported by Miloslav Vilimek (2008). The **tendon force** was calibrated using Bragg wavelength measurement of the FBG bonded on the tendon with applied load. FBG was used as displacement sensor on cadaver Achilles tendon and knee ligament for **movement measurement of tendons and ligaments** (Ren et al 2007). Study of change in length of ligament under various strain conditions is important as ligaments experience much higher strain as compared to bone for same loading. The FBG sensors exhibited higher sensitivity, low noise and same accuracy as compared to stereo-optic measurement which are though non-invasive have limitations of poor accuracy and high noise level.

In a very recent development a research group from Portugal has investigated osteoblastic **biocompatibility** of optical fibers and stability of the properties of FBG sensors for their in-vivo usage. (Carvalho et al 2011) The study analysed the behaviour of human bone marrow cells cultured in osteogenic-induced conditions over an optical fiber and in parallel, the sensing capability of FBG sensors throughout the culture time was assessed. The results showed that in addition to the excellent osteoblastic cytocompatibility, FBGs maintained the physical integrity and functionality, as their sensing capability was not affected throughout the cell culture period. Results suggest the possibility of in-vivo osseointegration of the optical fiber/FBGs anticipating a variety of applications in bone mechanical dynamics.

6.2 Biosensors

The ability of LPGs to detect refractive index variation in their vicinity has great potential for detection of clinical analytes and can be made to detect extremely low concentrations. An LPG with an immobilized antibody film on its surface is a very efficient device to detect target antigen bonding to this film by means of refractive index change associated with the process. The advantage of using LPG is that it is a direct and label free sensor which does not require any additional reagents to visualize binding. Figure 6 indicates the basic experimental set up for an LPG based biosensor. DeLisa et al (2000) have first reported use of LPG as biosensor for detection of human IgG by specific antibody-antigen binding with immobilized goat anti-human IgG antibody on the chemically treated surface of LPG. The system could work for antigen solution concentration between 2-100 $\mu g\ mL^{-1}$

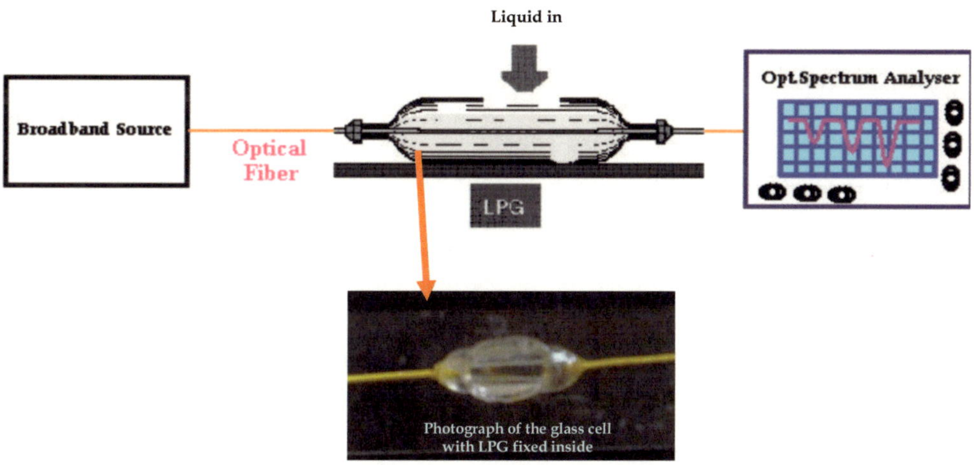

Fig. 6. The Experimental set up for an LPG Based Biosensor

Luna Analytics Inc. (Blacksburg, VA) had recently started developing an LPG based biosensor system though the product is yet to be commercialized. The sensitivities of these LPG sensors were found to be comparable to those of ELISA techniques (Baird & Myszka 2001, Pennington et al, 2001). LPG sensors have also been used for monitoring microbial activity (Carville, 2002]. Higher sensitivity in LPG sensors can be achieved by using gratings with smaller period or reduced- diameter cladding (Patrick & Kersey, 1998, Shu et al 2002). Chen et al (2007) have verified high sensitivity of smaller period LPGs and their reusability by detecting DNA hybridization.

In normal FBGs as the optical signals are confined to the fiber core regions they are insensitive to external refractive index variation, but they can be made sensitive either by writing tilted grating or etching the cladding part. A Bovine Serum Albumin (BSA) immobilized tilted FBG based immunosensor was reported recently to detect anti-BSA (Maguis et al 2008). In another approach, highly sensitive etched FBG sensors have been demonstrated by Chryssis et al (2005) to detect hybridization of single strand DNA. All the FBG based biosensing research is currently at the level of laboratory investigations only as it requires either specially designed or etched FBGs that are difficult if not impractical to fabricate and expensive as compared to LPGs.

FBG/LPG sensor systems can be incorporated with the microchips to detect chemical changes in nanolitre amounts of liquids and have the potential of being part of a **μ-TAS** (micro total analysis system) or **lab-on-chip**. Miniaturization and integration of light sources, sensors, detectors, as well as the corresponding signal processing is required for implementation of these concepts in a practical analysis system.

6.3 Cardiology

Flow-directed thermodilution catheters with conventional thermistor and thermocouple devices were commercially available for many years to monitor heart efficiency. However, as these sensors are electrically active they are not appropriate for use in a number of medical applications, especially in high magnetic fields associated with NMR machines. In a study, Rao et al (1997) and coworkers in UK used a catheter with FBG temperature sensor in place of thermocouple in a test rig set up to simulate blood flow in the heart with a peristaltic pump to simulate the heart pump. The results were found to be in good agreement with the electrical sensor with smoother temperature profile.

In 2005, Deniz Gurkan et al had proposed FBG sensor to monitor heartbeat using sound pressure for possible application in ballistocardiography (BCG). For a proof-of-concept demonstration, an FBG sensor was glued to vibrating membrane of a subwoofer of a speaker set. Using recordings of various heartbeat sounds, spectral changes were monitored and analyzed to extract all relevant information. They predicted that in a real life scenario, the FBG would be placed on the patient's body in the same way as a stethoscope to detect any abnormalities in the heart muscle more efficiently. FBG for blood pressure monitoring had been reported by Brakel et al (2004) where they had proposed FBG Fabry-Perot interferometer (FBGI) as a sensor configuration to detect strain resulting from blood pressure applied to the walls of an artery situated near the patient's skin. In an investigation they demonstrated an optical blood pressure manometer that not only measured accurate systolic and diastolic blood pressures once it was calibrated, but also provided a continuous pressure waveform quite comparable with conventional Sphygmomanometer pulse wave velocity system readings.

FBG sensors, being small and light, have particular advantages for use in endoscopic applications and offer the possibility of being used in conjunction with MRI scanning, thus opening up an opportunity for performing surgery with continual scanning. FBGs can be mounted on the tip of an intra-aortic catheter that serves as a laser-ablation delivery probe for the treatment of atrial fibrillation. The FBGs can give real-time, objective measure of tip-to-tissue contact force during the catheter ablation procedure. Force control is essential for delivering appropriate laser ablation pulses needed to produce lesions that are induced in the heart walls. If the contact force is too great, the catheter tip may perforate the heart wall and if it is too light, the procedure may be ineffective. Such force-sensing catheters with accompanying system have undergone extensive pre-clinical and clinical validation in the United States & Europe and are currently undergoing precertification trials[5].

6.4 Gynecology

Measurement of the pelvic muscles pressure was demonstrated using FBG based intravaginal probe in Portugal (Ferreira et al 2006). This measurement is essential for understanding pelvic floor disorders pathophysiology. The system was tested in a sample of patients with known pelvic floor disorders and the preliminary investigations indicated good sensitivity to radial pressure changes within the pelvic floor due to normal breathing cycle of the patient.

6.5 High intensity ultrasonic field measurements

High-intensity ultrasonic fields are used in various medical applications like ultrasound surgery, hyperthermia, lithotripsy and even diagnostic ultrasound. Thus safety concerns demand their accurate level assessment. The conventional detection techniques utilizing piezoelectric devices are susceptible to electromagnetic interference and signal distortion. Fisher et al (1998) demonstrated that FBG sensors can be used for this purpose by implementing them to detect signals of frequencies in the range of 500 kHz and 4 MHz.

6.6 Respiration monitoring system

A temperature independent FBG or LPG can be designed for non-invasive measurement of the torso movement during respiration ventilatory movements to understand respiratory physiology and to monitor the lung function. In a preliminary work, Günther Wehrle et al in 2001 employed FBG sensor on the chest using an elastic belt to hold it in place to detect thorax movements during artificial ventilation, even in the presence of electrical bursts caused by electrodes situated on the chest. Expansion of the thorax cage during respiration was accordingly transmitted to the sensor grating and caused it to deform under the strain. First application of a multiplexed LPG array on curvature sensing garment used to monitor the thoracic and abdominal movements of a human during respiration was reported by T. Allsop et al 2003, 2007. They have shown that, it is possible to generate a geometric profile of the chest and abdomen in three dimensions with an array of 20 sensors.

6.7 Temperature monitoring

Thermal therapy involves destruction of redundant tissue by heating or freezing without surgery. Examples of thermal therapies include treatment of benign prostatic diseases, ablation therapy of cardiac arrhythmia, microwave induced hyperthermia for radio therapy

---

[5] http://www.endosense.com/home.html

in the cancer treatment, during which temperature of both diseased and normal tissues should ideally be monitored. Most temperature sensors measure temperature at one point only, whereas FBG sensors because of their multiplexing capability provide temperature distribution during thermal treatment using one probe only. Temperature monitoring of a patient during MRI procedure is crucial in general and in particular to ensure that the implantable devices do not heat up under strong fields of MRI. An FBG probe can be placed inside a NMR machine with high magnetic field for remote temperature measurement.

An FBG based temperature profile monitoring system was first proposed by researchers from UK and Australia using four strain-free FBG array that meets the established medical requirements and tested the feasibility of FBG temperature probe inside a NMR machine with high magnetic field and established its working (Rao et al 1997, 1&2). The first *in-vivo* trials of such temperature probes incorporating five FBGs along a single fiber were undertaken successfully at the Cancer Research Institute in Perth, Australia by the same group (Webb et al 2000) on rabbits undergoing hyperthermia treatment of the kidney and liver via inductive heating of metallic implants. A distributed FBG sensor system was tested effectively in temperature range of -195.8^{0}C to100^{o}C for in-vivo use during freezing of porcine liver, for their mechanical stability and MR compatibility (Samset et al 2001).

FBGs are very useful in cryosurgery for localized cancer treatment, where precision of millimeters matters. It is very critical to monitor exact temperature profile throughout surgical procedure to ensure adequate freezing (-40°C) in the cancerous region but preventing low temperatures in the attached nerve fibers, blood vessels or nearby organs for their safety. An FBG based re-usable multi-point temperature monitoring system was developed recently which can record temperatures in a continuum of either four points at 10-mm intervals, or eight points at 5-mm intervals (Gowardhan & Greene 2007, Polascik & Mouraviev 2008).

6.8 FBG based manometry catheter

A research group from Australia has demonstrated multi-element catheter using FBG pressure sensor in in-vivo trials in the human oesophagus for diagnosis of gastrointestinal motility disorders (Arkwright et al 2009). The FBG based catheters had an outer diameter of less than 3 mm and were very flexible in comparison to solid state catheters, allowing easy nasal intubation and relative comfort for the patient during subsequent clinical diagnosis. Also, due to multiplexing capability of FBGs a large number of sensors can be accommodated in a single fiber thus making it possible to extend the total span of the sensing region, with no change in the catheter size. For validation, 10 healthy volunteers were intubated via the nose with both catheter assemblies and asked to perform series of controlled water swallows. The results were directly compared with a commercially available solid state catheter and largely equivalent response was obtained.

6.9 Smart clothing

Smart clothings have the capability to sense stimuli from the environment, to react and adapt to them. Due to their small size and flexibility fiber optic sensors can be woven into the textile and be an integral part of the smart system. This type of wearable and smart textile is useful not only for patient health monitoring but also for soldiers in battlefields, people in sports and even for general safety of public. In a collaborative work 10 expert groups from 5 EU countries have initiated a novel project, i.e. *Optical Fiber Sensors Embedded into Technical Textile for Healthcare* (OFSETH) to create smart wearable textiles with FBG along with other optical devices for measurements of various vital parameters[6]. Though preliminary experiments with the textile containing FBGs had been undertaken further investigations are needed for practical implementation of smart textile concept in terms of better integration of the fiber with textile, of different patient profiles and overall performance in real environment. (Grillet et al 2008)

6.10 Robotic surgery

A force sensing robot fingers using embedded FBG sensors and shape deposition manufacturing have been demonstrated recently (Park et al 2008). With their very small size and highly precise real-time measurements FBG sensors have the potential to be a part of surgical robot. Shape-sensing systems can be created by using arrays of FBGs distributed along multicore, singlemode fibers that will help determine the precise position and shape of medical tools and robotic arms used during minimally invasive/robotic surgery.

6.11 Optical Coherence Tomography (OCT)

OCT is an upcoming technology for non-invasive cross-sectional imaging in biological systems with a great potential for morphological assessment of various tissues. OCT acquires very high-resolution images through the detection of backscattered near infrared light with the potential to identify a wide range of pathologies in-vivo. Clinical applications of OCT are constrained by its limited penetration depth in tissue and miniaturized fiber-optic probes are used to image deeper within the body. (McLaughlin & Sampson 2010)

FBGs with precise Bragg wavelength, narrow spectral bandwidth less than 0.1 nm and high reflectivity are being applied in the OCT systems to calibrate and align the custom-made spectrometer for the spectral-domain optical coherence tomography (SD-OCT) system operating in the 1300-nm wavelength range. The calibration and characterization protocol was presented and investigated very recently. (Eom et al 2011)

7. Conclusion

Fiber grating based sensors and devices are just beginning to tap the vast opportunity for diagnostics to help in therapeutics as indicated by the large number of ongoing research & development activities mentioned in the previous section. They are enabling acquisition of analytical results on the patient's bedside within a few minutes minus the electromagnetic or other interferences. With the development of technology the possibility of cheaper and portable interrogator systems is within reach and the cost of grating fabrication would come down for bulk fabrication. Also, acceptance by the end users will follow only after the researchers are able to prove that this technology is better than the existing ones through

[6] http://www.ofseth.org

intensive collaborative efforts. In view of these aspects, it can be predicted that this rapidly evolving technology shall provide much better and effective solutions for various biomedical applications and has the potential to be an integral part of the futuristic healthcare systems.

8. Acknowledgment

The authors acknowledge Dr Pawan Kapur, Director CSIR-CSIO for his motivation and encouragement. They thank Mr SC Jain, Mr Umesh Tiwari and Mr GC Poddar from CSIR-CSIO, Chandigarh for their help and cooperation.

9. References

Allsop T., Earthrowl-Gould T., Webb D. J. & Bennion I. (2003) Embedded progressive-three-layered fiber long-period gratings for respiratory monitoring, *J Biomed Opt.* vol.8(3): 552–558

Allsop T., Carroll K., Lloyd G., Webb D.J., Miller M., Bennion I. (2007) Application of long-period-grating sensors to respiratory plethysmography, *J Biomed Opt.* vol.12 (6): 064003.

Arkwright J.W., Blenman N.G., Underhill I.D., Maunder S.A., Szczesniak M.M., Dinning P.G. & Cook I.J. (2009) In-vivo demonstration of a high resolution optical fiber manometry catheter for diagnosis of gastrointestinal motility disorders, *Opt Express*, vol.17(6): 4500-4508.

Baird C.L. & Myszka D.G. (2001) Current and emerging commercial optical biosensors, *J. Mol. Recognit.* vol.14: 261–268

Brakel A.V, Swart P.L., Chtcherbakov A.A. & Shlyagin M.G.(2004) Blood pressure manometer using a twin Bragg grating Fabry-Perot interferometer, *Photonics Asia Conference on Advanced Sensor Systems and Applications II*, Beijing, *Proc. SPIE* vol.5634:595-602

Carvalho L., Alberto N.J., Gomes P.S., Nogueira R.N., Pinto J.L., Fernandes M.H. (2011) In the trail of a new bio-sensor for measuring strain in bone: osteoblastic biocompatibility, *Biosens Bioelectron.* vol. 26(10): 4046-52

Carville D.G.M. (Jan 2002) Fiber optics for the detection of clinical analytes, *IVD Technology.* http://www.ivdtechnology.com/article/fiber-optics-detection-clinical-analytes

Chen X., Zhang L., Zhou K., Davies E., Sugden K., Bennion I., Hughes M. & Hine A. (2007) Real-time detection of DNA interactions with long period fiber-grating-based biosensor, Opt Lett, vol. 32(17): 2541.

Chryssis A.N., Saini S.S., Lee S.M., Yi H., Bentley W.E. & Dagenais M. (2005) Detecting Hybridization of DNA by Highly Sensitive Evanescent Field Etched Core Fiber Bragg Grating Sensors, *IEEE J. Sel. Top. Quantum Electron*, vol.11 (4): 864-872.

Cowie B.M., Webb D.J., Tam B., Slack P. & Brett P.N. (2006) Distributive Tactile Sensing using Fibre Bragg Grating Sensors for Biomedical Applications, *Proc. SPIE* vol. 6191: 249.

Cowie B.M., Webb D.J., Tam B., Slack P. & Brett P.N. (2007) Fibre Bragg Grating Sensors for distributive Tactile Sensing, *Meas Sci Technol*, vol.18: 138-146.

Davis, S. D., Gibbons, D. F., Martin, R. L., Levitt, S. R., Smith, J. and Harrington, R. V. (1972) Biocompatibility of ceramic implants in soft tissue, *Journal of Biomedical Materials Research*, vol. 6: 425–449

DeLisa M.P., Zhang Z., Shiloach M., Pilevar S., Davis C.C., Sirkis J.S. & Bentley W.E. (2000) Evanescent Wave Long-Period Fiber Bragg Grating as an Immobilized Antibody Biosensor. *Anal. Chem.*, vol.72: 2895-2900

Dennison C.R., Wild P.M., Wilson D.R., Cripton P.A. & Dvorak M.F.S. Pressure Sensor for Biological Fluids and use thereof, Patent application: WO 2007/09/5752.

Dennison C. R., Wild P. M., Byrnes P.W.G., Saari A., Itshayek E., Wilson D. C., Zhu Q. A., Dvorak M.F.S., Cripton P.A.& Wilson D.R. (2008) Ex vivo measurement of lumbar intervertebral disc pressure using fiber-Bragg gratings, *J Biomech*, vol.41(1): 221-225.

Dennison C.R., Wild P.M., Wilson D.R., Cripton P.A. (2008) A minimally invasive in-fiber Bragg grating sensor for intervertebral disc pressure measurements, *Meas. Sci. Technol*, vol.19 :085201

Eom T.J., Ahn Y.C., Kim C. S. & Chen Z. (2011) Calibration and characterization protocol for spectral-domain optical coherence tomography using fiber Bragg gratings, *J. Biomed. Opt.* vol.16(3): 030501 1-3

Ferreira L.A., Araújo F.M., Mascarenhas T., Natal Jorge R.M., Fernandes A.A. (2006) Dynamic assessment of women pelvic floor function by using a fiber Bragg grating sensor system, *Proc. SPIE* vol. 6083: 60830H-1.

Fisher N.E., Webb D.J., Pannell C.N., Jackson D.A., Gavrilov L.R., Hand J.W., Zhang L.& Bennion I. (1998) Ultrasonic hydrophone based on short in-fiber Bragg gratings, *Appl Opt.* vol. 37 (34): 8120-28.

Fresvig T., Ludvigsen P., Steen H. & Reikeras O. (2008) Fiber optic Bragg grating sensors: An alternative method to strain gages for measuring deformation in bone, *Med Eng Phys* vol. 30: 104–108

Gowardhan B. & Greene D. (2007) Cryotherapy for the prostate: an *in vitro* and clinical study of two new developments; advanced cryoneedles and a temperature monitoring system, *BJU Int.* vol.100(2): 295-302

Grillet A., Kinet D., Witt J., Schukar M., Krebber K., Pirotte F. & Depré A. (2008) Optical Fiber Sensors Embedded Into Medical Textiles for Healthcare Monitoring, *IEEE Sensors J*,vol.8: 1215.

Gurkan D., Starodubov D. & Yuan X. (Nov. 2005) Monitoring of the heartbeat sounds using an optical fiber Bragg grating sensor, in *Proc. 4th IEEE Conf. Sensors*, Irvine, CA,

Hao J.Z., Tan K.M., Tjin S.C., Liaw C.Y., RoyChaudhuri P., Cuo X. & Lu C., (2003) Design of a foot-pressure monitoring transducer for diabetic patients based on FBG sensors, *LEOS 2003. The 16th Annual Meeting of the IEEE*, 23-24.

Hao, J., Jayachandran, M., Kng, P., Foo, S., Aung Aung, P., and Cai, Z. (2010). FBG-based smart bed system for healthcare applications. *Frontiers of Optoelectronics in China 3*, 78-83.

Hecht J. (1999) *City of Light: The Story of Fiber Optics*, Oxford University Press, New York,

James S.W. & Tatam. R.P. (2003) Optical fibre long-period grating sensors: Characteristics and application, *Meas. Sci. Technol.* vol.14: R49-R61

Kashyap R. Fiber Bragg Gratings, second edition (Academic Press, London), 2009.

Katzir A. (1989) Optical Fibers in Medicine, *Scientific American*, 120

Maguis S., Laffont G., Ferdinand P., Carbonnier B., Kham K., Mekhalif T. & Millot M.C. (2008) Biofunctionalized tilted Fiber Bragg Gratings for label-free immunosensing, *Opt Express*, vol.16(23) :19049.

Majumder M., Gangopadhyay T.K., Chakraborty A.K., Dasgupta K.& Bhattacharya D.K. (2008) Fibre Bragg gratings in structural health monitoring—Present status and applications, *Sens. Actuators, A* vol.147: 150–164.

McLaughlin R.A. & Sampson D.D. (2010) Clinical applications of fiber-optic probes in optical coherence tomography, *Optical Fiber Technology* vol.16: 467–475

Milgrom C., Finestone A., Simkin A., Ekenman I., Mendelson S., Millgram M., Nyska M., Larsson E. & Burr D. (2000) In vivo strain measurements to evaluate the strengthening potential of exercises on the tibial bone, *J Bone Joint Surg* vol. 82-B: 591-4.

Milgrom C., Finestone A., Segev S., Olin C., Arndt T. & Ekenman I. (2003) Are overground or treadmill runners more likely to sustain tibial stress fracture?, *Br J Sports Med* vol.37: 160–3.

Miloslav V. (2008) Using a fiber Bragg grating sensor for Tendon force measurements, *J Biomech*, 41(S1).

Mishra V., Singh N.& Kapur P. (2009) Fiber Optic Sensors Technology in Medicine, in: D.V. Rai, Raj Bahadur (Eds.), *Trends in Medical Physics and Bio medical Instrumentation*, New Era International Imprint, India, pp. 225-242

Mishra V., Singh N., Jain S. C., Kaur P., Luthra R., Singla H., Jindal V. K. & Bajpai R. P. (2005) Refractive index and concentration sensing of solutions using mechanically induced long period grating (LPG) pair, *Opt Eng*, vol. 44(9): 094402-1-4.

Mishra V., Singh N., Rai D.V., Tiwari U., Poddar G.C., Jain S.C., Mondal S.K. & Kapur P. (2010) Fiber Bragg Grating Sensor for Monitoring Bone Decalcification, *Orthop Traumatol Surg Res*, vol. 96: 646- 651

Mishra V., Singh N., Tiwari U.& Kapur P.(2011) Fiber Grating Sensors in Medicine: Current and Emerging Application, *Sens. Actuators, A*. vol.167: 279–290,

Mohanty L., Tjin S.C., Lie D., Panganiban S.E.C., Chow P. (2007) Fiber grating sensor for pressure mapping during total knee arthroplasty, *Sens. Actuators, A*. vol.135(2): 323-328

Othonos A. & Kalli K. (1999) *Fiber Bragg Gratings: Fundamentals and Applications in Telecommunications and Sensing*, Artech House, London.

Ottevaere H., Tabak M., Fernandez A.F., Berghmans F.& Thienpont H. (2005) Optical fiber sensors and their application in monitoring stress build-up in dental resin cements, *Optical Fibers: Applications, Warsaw, Proc. SPIE, v*ol. 5952: 204-216,

Patrick H.J. & Kersey A.D. (1998) Analysis of the response of long period fiber gratings to external index of refraction, *J. Lightwave Technol*, vol.16: 1606.

Park Y.L., Ryu S.C., Black R.J., Moslehi B. & Cutkosky M.R. (2008) Fingertip Force Control with Embedded Fiber Bragg Grating Sensors, *Proc IEEE International Conference on Robotics and Automation*, 3431 - 3436

Paul J., Zhao L., & Ngoi B.K.A. (2005) Fiber-optic sensor for handgrip-strength monitoring: conception and design. *Appl Opt*, vol.44 (18): 3696-3704.

Pennington C., Jones M., Evans M., VanTassell R. & Averett J. (2001) Fiber-optic-based biosensors utilizing long period grating (LPG) technology, *Proc. SPIE* vol. 4255: 53.

Pleros N., Kanellos G.T. & Papaioannou G. (2009) Optical Fiber Sensors in Orthopedic Biomechanics and Rehabilitation, *Proc. 9th International Conference on Information Technology and Applications in Biomedicine, ITAB 2009*, Cyprus, 1-4.

Polascik T.J. & Mouraviev V. (2008) Focal therapy for prostate cancer. *Curr Opin Urol* vol.18: 269–274.

Rao Y.J. (1997) In-fibre Bragg grating sensors, *Meas. Sci. Technol*, vol. 8: 355-375

Rao Y.J., Jackson D.A., Webb D.J., Zhang L. & Bennion I. (1997) In-fiber Bragg grating flow-directed thermodilution catheter for cardiac monitoring, *Proceedings of the Optical Fiber Sensors Conference (OFS-12)*, Williamsburg, VA, USA 354-357.

Rao Y.J., Hurle B., Webb D.J., Jackson D.A., Zhang L., Bennion I. (1997) In-situ temperature monitoring in NMR machines with a prototype in-fiber Bragg grating sensor system, *Proceedings of the Optical Fiber Sensors Conference (OFS-12)*, Williamsburg, VA, USA, 646-649.

Rao Y.J., Webb D.J., Jackson D.A., Zhang L., Bennion I. (1997) In-fiber Bragg grating temperature sensor for medical applications, *J. Lightwave Technol*, vol.15(5): 779-785.

Ren L., Song G., Conditt M., Noble P. & Li H. (2007) Fiber Bragg grating displacement sensor for movement measurement of tendons and ligaments, *Appl. Opt.* vol.46: 6867-6871

Russell P. S. J. (2003) Photonic crystal fibers, *Science,* 299 358–362.

Samset E., Mala, T. Ellingsen R., Gladhaug I., Søreide O. & Fosse E. (2001) Temperature measurement in soft tissue using a distributed fibre Bragg grating sensor system. *Min Invas Ther & Allied Technol*, vol.10(2): 89–93

Schiller M. W., I. Abe, P. Carvalho, P. Lopes, L. Carvalho, R. N. Nogueira, J. L. Pinto, J. A. Simões, On the Use of FBG Sensors to Assess the Performance of a Dental Implant System, *Proceedings of the Optical Fiber Sensors Conference (OFS-2006)*, (Oct. 23-27 2006) Cancún, México. ThE77.

Shu X., Zhang L. &. Bennion I. (2002) Sensitivity Characteristics of Long-Period Fiber Gratings, *J. Lightwave Technol*, vol. 20(2): 255-266

Silva J.C.C., Carvalho L., Nogueira R.N., Simões J.A., Pinto J.L. & Kalinowski H.J. (2004) FBG applied in dynamic analysis of an implanted cadaveric mandible, *Second European Workshop on Optical Fibre Sensors (Santander, Spain), Proc. SPIE* 5502 : 226-229

Talaia P.M., Ramos A., Abe I., Schiller M.W., Lopes P., Nogueira R.N., Pinto J.L., Claramunt R. & Simões J.A. (2007) Plated and Intact Femur Strains in Fracture Fixation Using Fiber Bragg Gratings and Strain Gages, *Exp Mech*, vol.47: 355–363

Tjin S. C., Tan Y. K., Yow M., Lam Y. Z. & Hao J. (2001) Recording compliance of dental splint use in obstructive sleep apnoea patients by force and temperature modeling, *Med Biol Eng Comput*, vol.39:182-184

Tiwari U., Mishra V., Poddar G.C., Kesavan K., Jain S.C., Ravisankar K., Singh N. & Kapur P. (2009) Health Monitoring of Steel and Concrete Structures using Fiber Bragg Grating Sensors, *Current Science*, vol. 97(11) 1539-42.

Tiwari U., Mishra V., Bhalla A., Singh N., Jain S.C., Garg H., Raviprakash S., Grewal N. & Kapur P. (2011) Fiber Bragg grating Sensor for Measurement of Impact Absorption Capability of Mouthguards, Dental Traumatology, vol.27: 263-268.

Webb D.J., Hathaway M.W., Jackson D.A., Jones S., Zhang L., Bennion I. (2000) First in-vivo trials of a fiber Bragg grating based temperature profiling system, *J Biomed Opt*, vol.5: 45–50

Wehrle G., Nohama P., Kalinowski H., Torres P.I. & Valente L.C.G. (2001) A fiber optic Bragg grating strain sensor for monitoring ventilatory movements, *Meas. Sci. Technol.* vol.12: 805–809

Yang C., Zhao C., Wold L., Kaufman K.R. (2003) Biocompatibility of a physiological pressure sensor, *Biosensors and Bioelectronics* vol.19: 51-58

7

Additive Manufacturing Solutions for Improved Medical Implants

Vojislav Petrovic, Juan Vicente Haro, Jose Ramón Blasco and Luis Portolés
Metal-Processing Research Institute AIMME, Valencia
Spain

1. Introduction

In recent years, European industry has been facing the challenge of losing competitiveness in mass production. Due to important factors such as lower labour costs, lower taxes or in-site access to raw materials, mass production has migrated to Third World countries. However, European industry is more advanced in technological aspects and is in need of a qualitative advantage in the development of new technologies. One of the efforts of European companies is directed towards the production of short series of customized products with added value. Major efforts have been done in order to customize products and give them an added value by developing new manufacturing technologies.

Additive Manufacturing (AM) is a powerful tool that offers the necessary competitiveness to European companies (Petrovic, 2011). AM technologies have been available on the market for many years. Initially, these technologies were considered only for prototyping - the first technologies that appeared on the market were capable of fabricating only polymer parts of low quality and low resistance. However, in the last decade, the sector of AM has experienced an important evolution with constant growth in sales of machine systems and rapid products (Wohlers, 2010). Numerous advantages of 'freeform fabrication' have driven new developments in processing principles and materials. New value-added materials have been released for layer-by-layer processing. On the other hand, new technologies have been developed to process demanding materials for different sectors. New energy sources have been introduced in order to process high melting point metal alloys such as Titanium, Cobalt Chromium, etc.

There are many terms commonly used for AM, such as solid *free form fabrication* (FFF), *rapid manufacturing* (RM), *additive layered manufacturing* (ALM) and *3D printing*. The latter may be the most descriptive for people not familiar with additive technologies. Unfortunately, this term may produce a wrong idea, since AM machinery is much more than any kind of printer. However, officially and according to ASTM F42 Committee, Additive Manufacturing is defined as "process of joining materials to make objects from 3D model data, usually layer upon layer, as opposed to subtractive manufacturing methodologies, such as traditional machining" (ASTM, 2010).

AM enables the use of **value-added design** in medical device manufacturing sector. Process of adding material in layers allows the fabrication of *designed, controlled and well-*

interconnected porosity which, combined with solid material, provides better bone ingrowth into implants. Also, AM implants are characterized by *rough surface quality* per se. Undesirable in other sectors, in medical implants rough surface is an advantage because it enhances bone-implant fixing. Furthermore, AM technologies perform fabrication of metal implants in a highly controlled atmosphere with restricted presence of oxygen, which results in especially *high purity*. Finally, layer-by-layer principle allows the fabrication of *customized net-shape implants* that fully fit patient's data. In addition, high power and processing velocity open the possibility of serial production of *standardized implants*.

Fig. 1. Acetabular cup with designed porous surface (Courtesy of ARCAM AB).

The aim of this chapter is to illustrate the capabilities of AM on the example of a technology with one of the most powerful active principles: Electron Beam Melting (EBM).

2. Additive manufacturing process

Additive fabrication is performed directly from a 3D CAD file in which a geometrical model of part is stored. The part model can be designed in many different commercial 3D modelers but it is exported in STL file format. The STL file is imported into a specific software (such as Magics® by Materialise, Viscam® by Marcam, Netfabb® by FIT, etc.), where it is preprocessed. The part is oriented for building and a support structure is made for the downfacing surfaces of the part. Afterwards, cross-sections of a given thickness, known as 'slices', are generated virtually from 3D CAD descriptions of the part and support structures. The slices are saved in a 'slice file' (ABF, SLI, SLC or any other format that may depend on patented technology). After pre-processing, the slice file is sent to the machine to be 'printed' slice-by-slice.

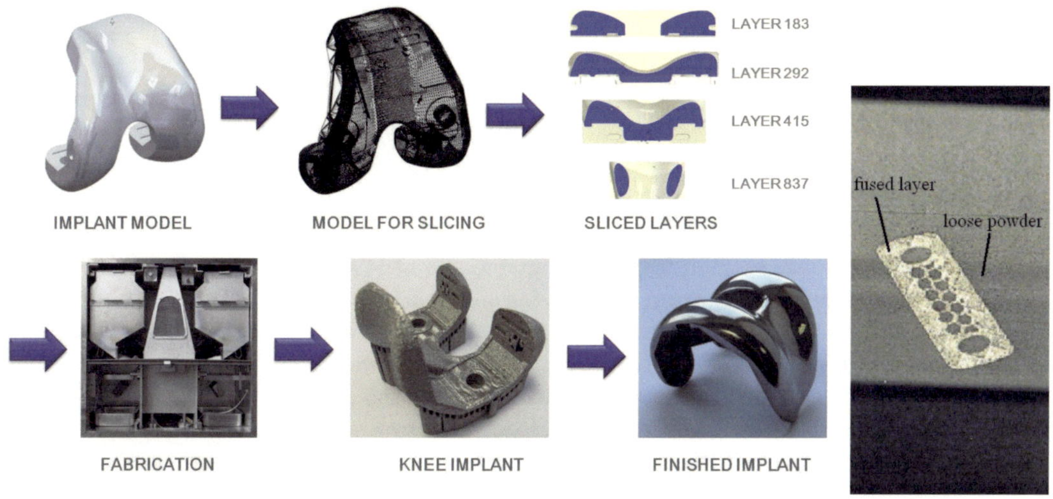

Fig. 2. Steps in AM process (left) and fused layer (right)

The fabrication process consists of two basic steps: *coating* and *selective melting*. The coating step is the process in which material is laid over the working surface in a very thin layer. The thickness of the layer depends on the chosen technology and it ranges between 0.03 mm to 0.20 mm. The selective melting step refers to the process of printing the part slice by the action of a source of energy. The active principle can be a *light source*, *laser beam*, *electron beam*, etc. It acts over the layer of material and produces fusion of slice's footprint to the layer below. The power of the energy source depends on the chosen technology: in the low end, we have Stereolithography that uses ~100 mW, and in the high end, we have Electron Beam Melting that uses up to 3000 W. Only the contour of part slice and its interior are fused (Figure 2, right); the rest of the material is left untouched and may be recycled (more or less, depending on the process and material). The actions of coating and slice printing follow each other until all the slices are correctly printed, generating the final three-dimensional part. Once fabrication has finished, the part is taken out to be cleaned of material that is not fused and, if necessary, to be given further post-processing. The most common post-processing techniques are sanding, polishing, homogenization, thermal treatment, etc.

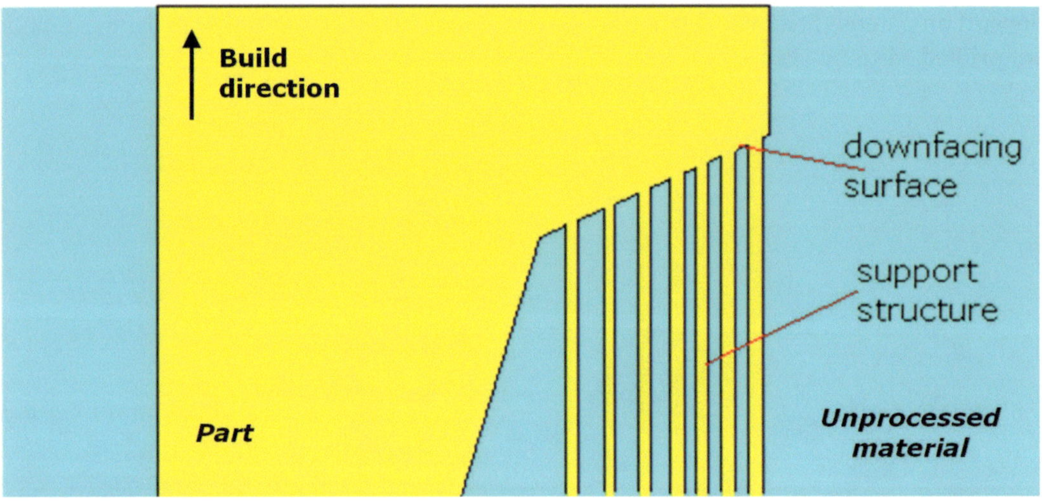

Fig. 3. Support structure for downfacing surfaces

It is important to highlight that, in general, material that is not processed (be it powder or liquid) is not capable of supporting fused material. Hence, if a part has overhanging zones, they may need to be supported by a *support structure* (Figure 3). Additionally, the support structure acts as a conductor of excess heat created in the process of selective melting. Finally, in some technologies, the support structure prevents warping of the part due to the thermal stress created during the process. Hence, the first step of pre-processing actually consists in generating an efficient support structure, which is then sliced and fabricated together with the part. The support structure is eliminated in the post-processing.

The most important advantages of Additive Manufacturing are:

1. *Time-to-market reduction for customized products.* Due to high speed of the process and being a direct fabrication method, the reconstruction of a model to fit in desired assembly is relatively fast;
2. *Product customization with complete flexibility in design & construction of a product.* Unlike conventional processes, AM can produce parts of almost any desired form and can almost be free from geometrical manufacturing constraints.
3. *Maximum material savings.* Material is added and not subtracted. For some applications, especially in the metal sector, case studies show that the waste of raw material is reduced by up to 40% when using additive technologies instead of subtractive (machining) technologies (Reeves, 2008). Also, 95% to 98% of the remaining material (powder or resin that is not processed) may be recycled;
4. *No tools, moulds or punches are needed.* The part is obtained directly from its 3-D CAD model with the absence of human errors in production.
5. *Full-density of final parts.* Unlike other powder based processes (powder metallurgy, MIM), additive technologies produce parts without almost no residual porosity.

6. *Fabrication of free form enclosed structures.* Additive technologies are capable of fabricating free form channels as well as different forms of latticed structures.

However, from the point of view of their application in biomedical field, additive technologies are facing some challenges:

1. *Remove the 'stigma' of its original name 'Rapid Prototyping'.* Although AM has evolved to deliver ready-to-use products made from a wide variety of metals and polymers, it is still wrongly considered that AM is valid only for prototype fabrication.
2. *Validation of mechanical properties of existing materials and AM technologies.* Unlike conventionally produced parts, AM parts may not behave identically in all directions. Depending on particular additive technology, processed material has better behaviour when the load is applied along the direction of the layer as compared to the build-up direction.
3. *Development and characterization of new materials for AM.* Alloys like stainless steel, titanium alloys, cobalt chromium, etc. are already being processed. Nevertheless, there are many interesting materials that are under research or considered for further research (see Future development section).

Electron Beam Melting

EBM is one of free-form fabrication technologies capable of processing ferrous and non ferrous metallic powders to fully-dense material, using layer-by-layer principle. In the case of EBM, the energy is delivered through an electric circuit of 60kV that is created between a tungsten filament placed inside of the electron gun and the building plate (Figure 4).

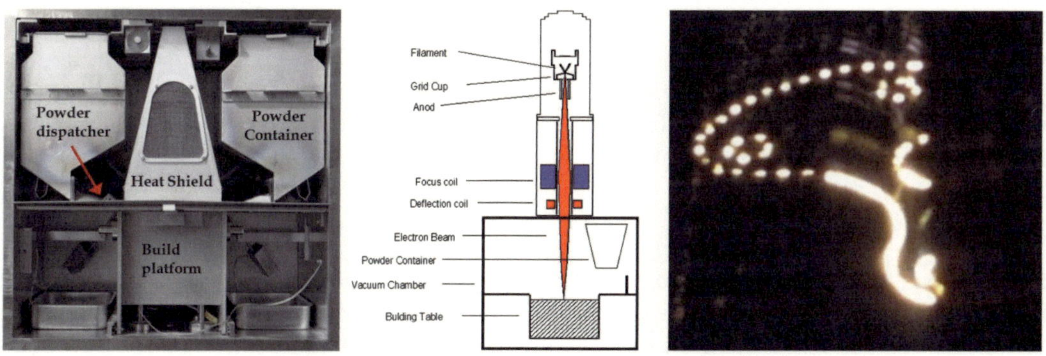

Fig. 4. Inside the EBM machine (left); scheme of an additive machine (middle); electron beam making the contour of tibia prosthesis.

The filament is heated by electric current and emits a beam of electrons which is conducted by a set of different coils until it impacts the powder surface. During the impact, electric energy is transformed to heat energy which fully melts the powder. The working chamber is kept under deep vacuum (order of magnitude 10^{-4} mbar). Hence, powder is released from containers and distributed over the build platform in fine 70-100 µm layers. The beam melts powder to a solid slice, following the cross-section of the part at that layer and merging it with previous slices (Figure 4). The build platform descends for the value of layer thickness and a new powder layer is dispatched. The process repeats until the part is completed.

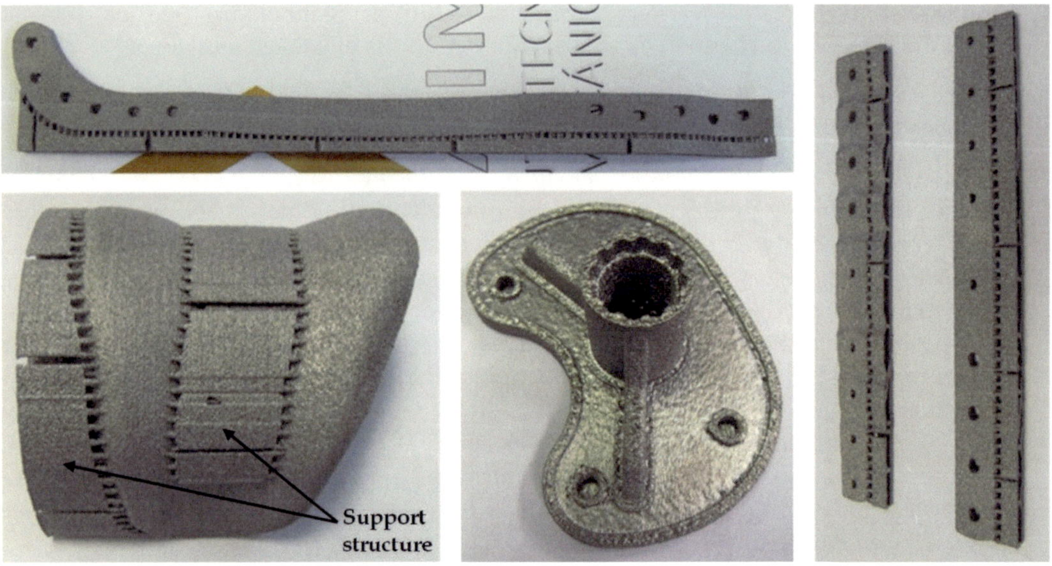

Fig. 5. Some of biomedical parts as produced on the machine (osteosynthesis plates Courtesy of CIMA).

In comparison with other AM processes, EBM has three major advantages very relevant for medical implants manufacturing:

- *Substantially higher nominal speed* which makes viable a serial production of medical implants (Figure 5).
- *Processing under vacuum* - the processed material has very high purity resulting in higher properties and better biocompatibility.
- *High processing temperature* (for Ti64 around 650°C) - less thermal stresses and less warpage in processed material.

3. Supply chain flow

Supply chain for medical implants fabricated by AM consists of six steps (Figure 6). Four of them belong to the fabrication process while the others are common to all medical device manufacturing processes (medical post-treatment and surgical intervention).

3.1 Implant reconstruction / design

As mentioned before, AM process can be understood as 3D printing of solid models. There are two different ways to obtain medical models:

- *Designing the model in some 3D modelling software.* Based on the statistic information about patients (age, weight, physical constitution, etc.), implants are designed together with tooling for the surgical operation (Synthes, 2007). This is a commonly used way to make standard implants in different sizes. When implanted, the osseous zone is adjusted so that the implant fits to properly repair the damaged zone. The model can be manufactured by a variety of processes (forging, CNC machining, etc.). However, additive technologies introduce one important advantage: porous surface for better bone ingrowth (Figure 7). This *controlled* porous coating is designed in a specific software (in the case shown on Figure 7, it was Magics® by Materialise) and exported as a model coupled with the solid hip stem.

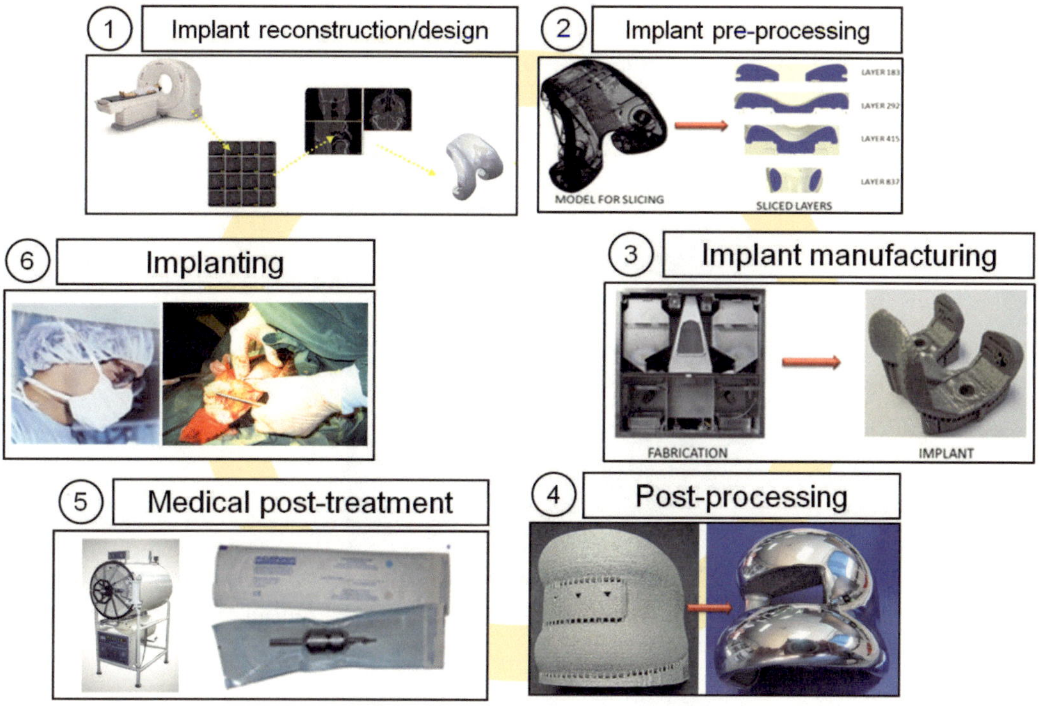

Fig. 6. Supply chain flow for medical implants.

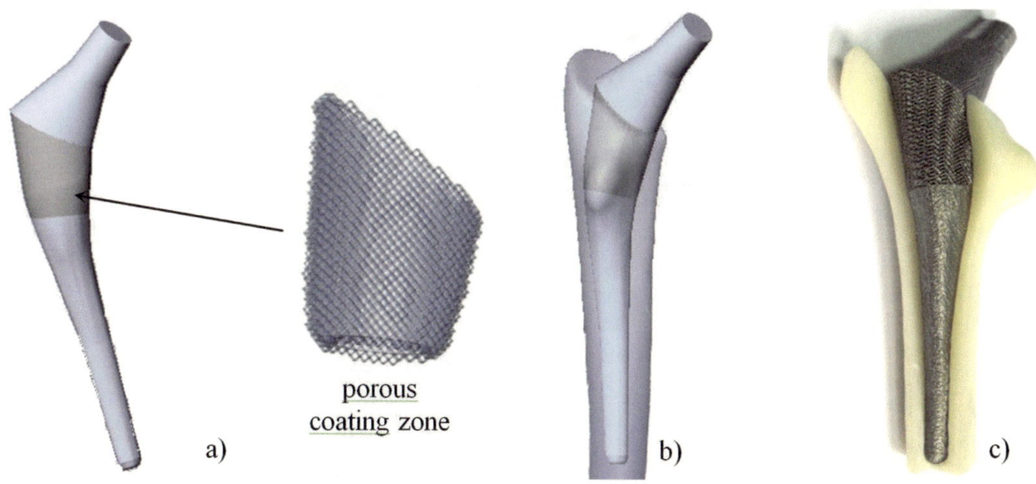

Fig. 7. Example of femoral hip stem: a) 3D model of hip stem with superficial porous zone, b) hip stem coupled with femur bone (model) and c) Ti64 hip stem coupled with polyacrilic bone replica.

- *Reconstruction of model upon patients CT images.* For customized implants, the common path is to reconstruct the model via *scan-to-part*: a cloud of points is reconstructed upon CT images and subsequently converted into a 3D model. The model can then be manufactured in chosen technology (Figure 8). As in each process of reverse engineering, there is an error that is introduced during the reconstruction. In the scan-to-part process, the maximum introduced deviation was 1.4 mm - the majority of the model points have deviation comprehended between 0.45 and 0.65 mm (Figure 9, left). On the other hand, the fabrication process reproduces the model with the deviation inferior to 0.15mm in more than 80% of points (Figure 9, right) (AIMME, 2009).

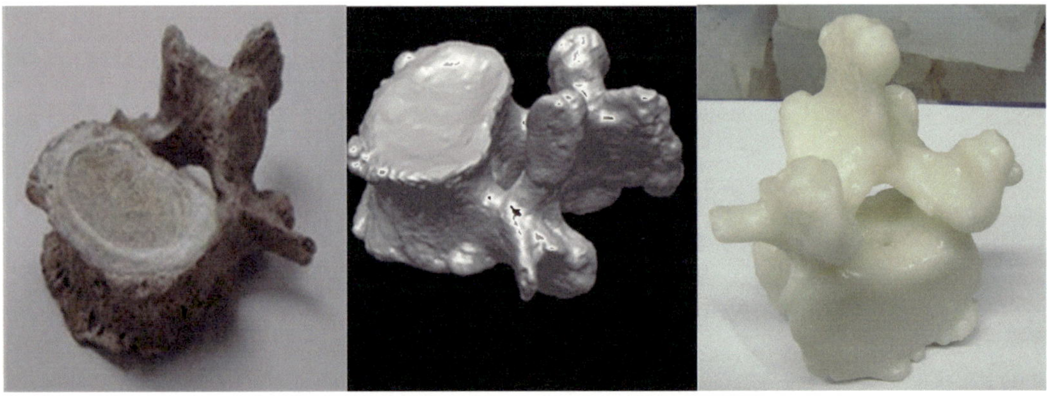

Fig. 8. Example of spinal vertebra: a) real bone, b) reconstructed model and c) polyacrilic bone replica made on Stereolithography.

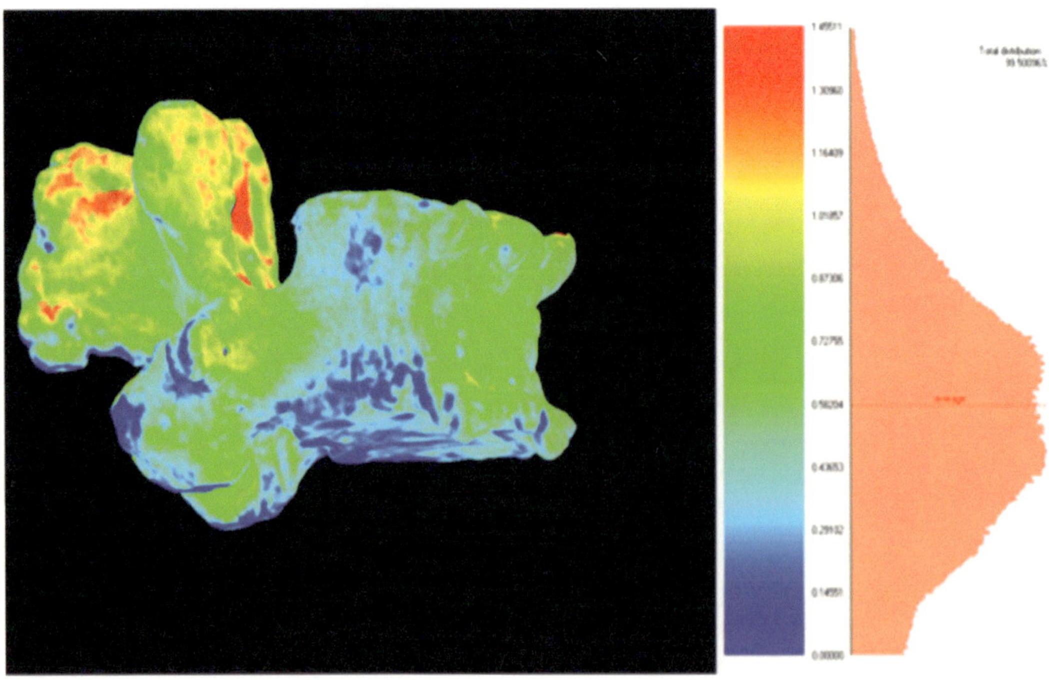

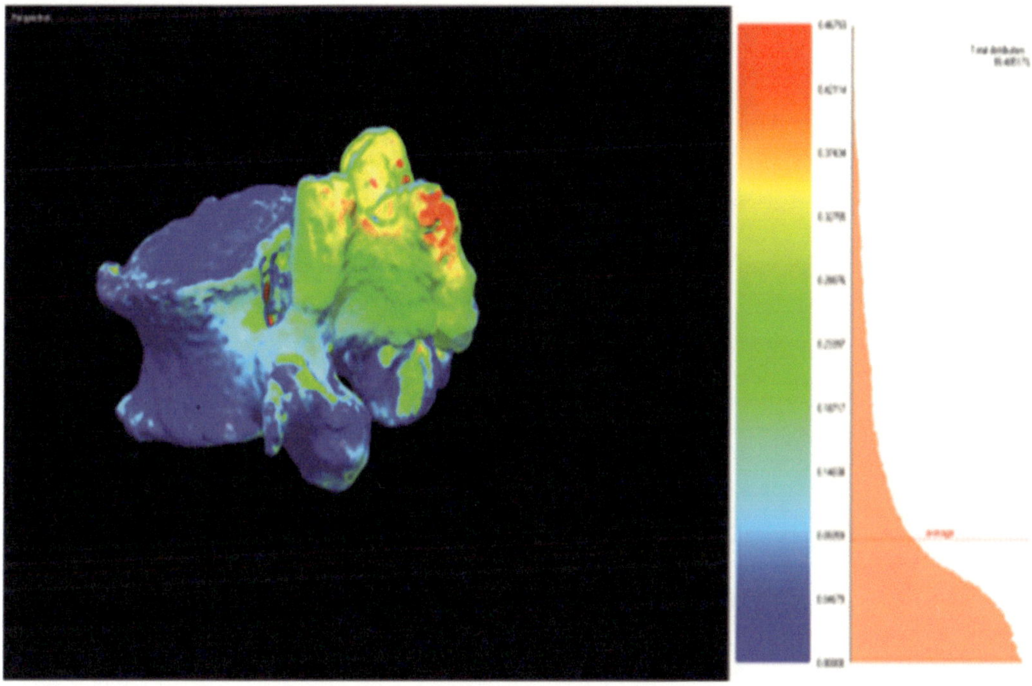

Fig. 9. Comparison by light digitalization: a) deviation of real vertebra vs reconstructed STL model, b) deviation of polyacrilic bone replica made on Stereolithography vs reconstructed STL model.

After completing the scan-to-part, in Additive Manufacturing the model is sliced virtually and then built layer by layer. This fact allows the fabrication of very complex shapes and forms. In biomedical applications, it allows the fabrication of near net-shape implants customized to the patient's data.

The same solutions of gradual porosity mentioned previously can be applied to customized implants.

3.2 Implant pre-processing

The planning of manufacturing process of implants in Additive Manufacturing consists of two steps:

- Implant model is *properly oriented on the build platform* for layer-by-layer fabrication in order to optimize surface quality, support structure, build time, build cost, etc. If necessary, the support structure is generated and optimized. As much implants as possible are packed for more efficient fabrication (Figure 10). For the time being this is done manually, but some tools for automatic assessment are being developed for knowledge assisted part orientation (Petrovic, 2010).
- *Implant and support structure are sliced for fabrication.* Slices are stored into a sliced file which is uploaded to the machine (Figure 11). The machine uses specific software to interpret the file and to send commands to print layers into a solid part. The same build, stored in the sliced file, can be build again without any pre-processing.

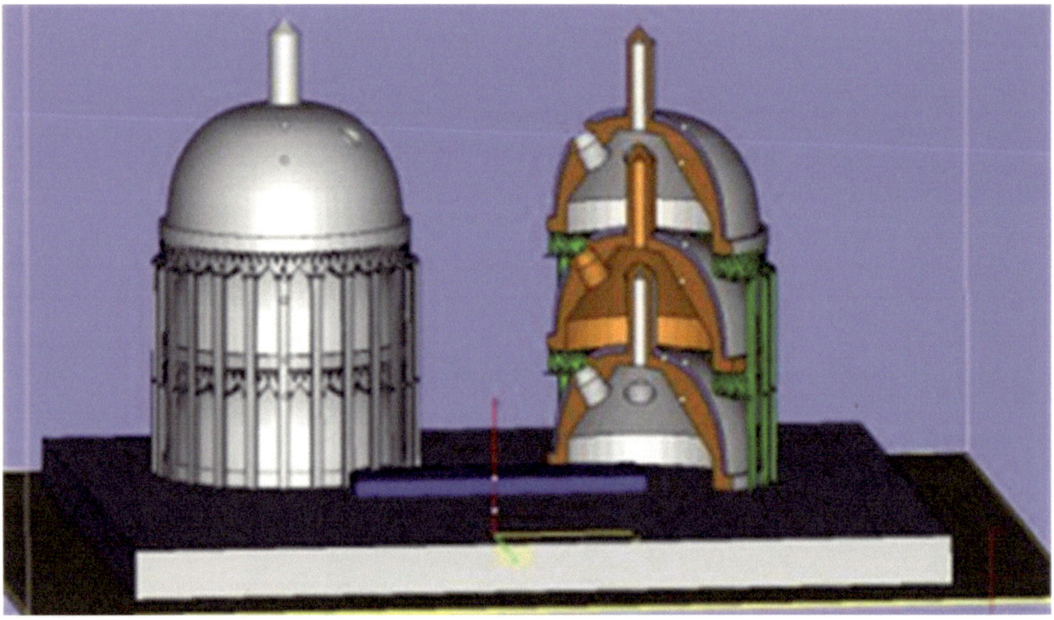

Fig. 10. Packing of implants for more efficient fabrication.

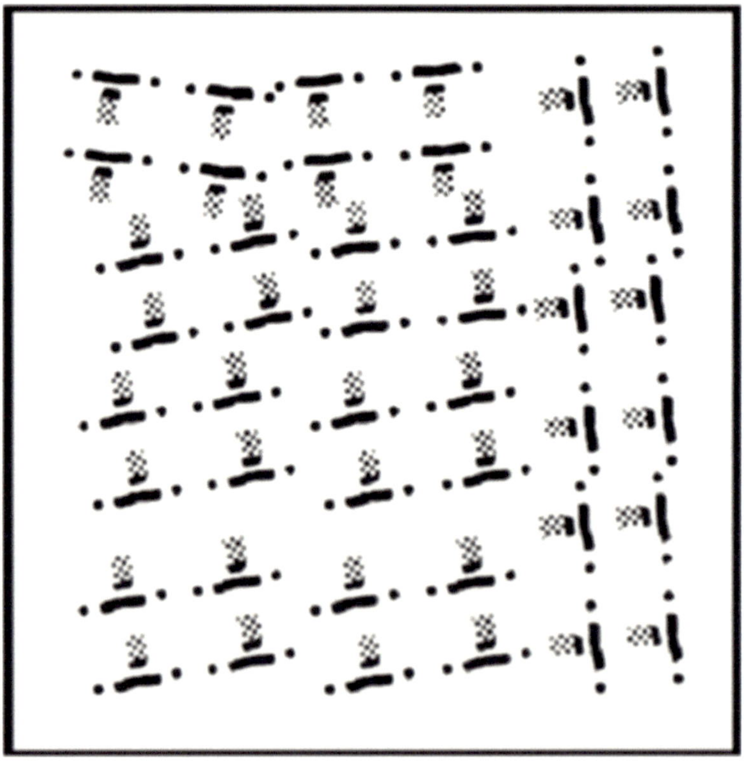

Fig. 11. One digital layer as represented in EBM Control software (black part is fused and white left unfused).

3.3 Implant fabrication

Once the pre-processing is done, implants are manufactured layer by layer on the machine. In the case of EBM, as mentioned before, the printing of the layers is done by the action of electron beam which performs a selective melting of material (fuses the black part on Figure 10 and leaves unfused the white part). The electron beam is very powerful and has a diameter of approximately 240 microns, which makes the contour of the layer - that is the surface of the implant when finished - a bit rough (Figure 12). This implies additional post-processing in the zones of the implant that need smooth surface. However, in order to get better surface quality, the technology has suffered some changes recently and a *Multibeam®* strategy has been introduced (Figure 12).

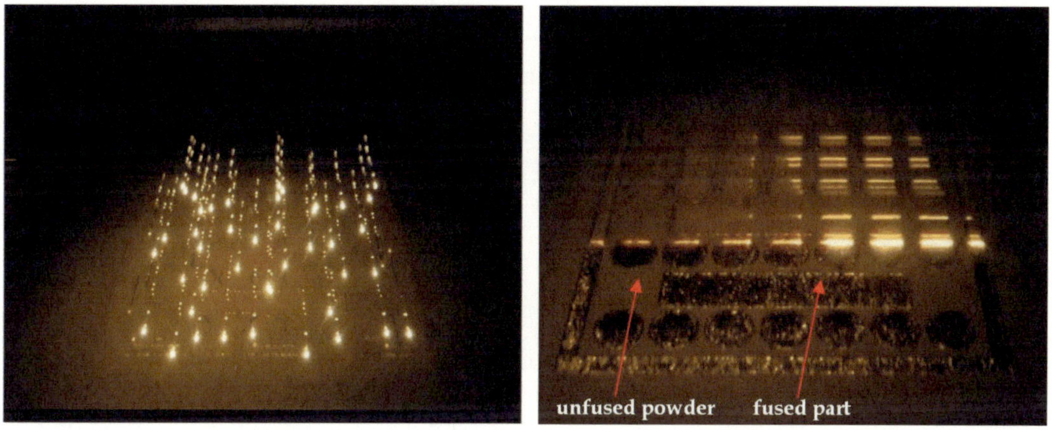

Fig. 12. Multibeam® strategy for better implant surface quality (left), fusing the layer (right)

3.4 Implant post-processing

After the fabrication has finished, the building platform with implants is taken out of the powder bed (Figure 13). Major part of the unfused powder is directly poured into a sifting system and filtered for reuse. However, in the case of EBM, the working temperature of the chamber reaches high value (650°C for Ti64). Hence, the unfused powder is semi-sintered which is why, in order to clean the implants thoroughly, they are transferred to *Powder*

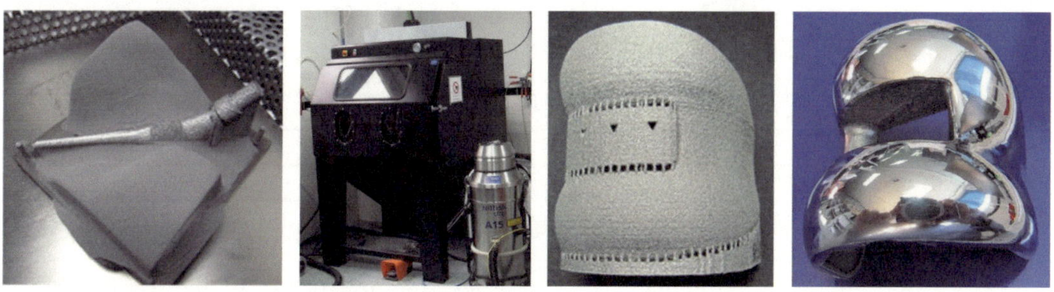

Fig. 13. Implant post-processing. From left to right: hip stem still involved by the powder bed; powder recovery system; knee implant as taken out from the machine and after support removal and hand polishing.

Recovery System (PRS). The rest of powder is wiped away by a jet of compressed air charged with particles of the same material that the implants are made of (so as to prevent implant contamination). After cleaning the implant, the support structure is removed by hand (it is designed to have small contact surface with implant and be easy to eliminate). Also, if necessary, additional machining and/or polishing of certain surface or zone is performed.

As can be seen from this analysis of the supply chain, main advantages of additive technologies are:

- *value-added design* (porous surface with controlled porosity),
- *use of additive rather than subtractive fabrication* which allows:
    - much bigger geometrical freedom;
    - recycling of major part of material (up to 98%);
- *automation* of the process, which permits to:
    - avoid human errors since the batch of models is stored in electronic way and prepared for "load & play";
    - make hundreds of implants in a week time with very optimized price, depending on size, since the same batch can be built again without additional preparation.

4. Processed material

The raw material in EBM is powder. Hence, it is reasonable to expect that the processed material contains residual porosity. However, the processed material on EBM is almost 100% dense because of complete local fusion of powder. Experimental results are offered to illustrate the capabilities of EBM in biomedical applications through two widely used materials: Ti6Al4V and CoCr ASTM F75 which are commercially available for processing on EBM. More materials are under development which will be mentioned in the 'Future Developments' section.

4.1 TiAl6V4 alloy

Ti6Al4V is a widely used biomaterial for many medical applications (Biomet, 2009), (Oshida, 2007), (Bronzino, 2006). Experimental results show that the properties of full-dense Ti64 processed on EBM fulfill the corresponding norm for medical implants (ISO, 2010), (ASTM, 2010) and are even superior to casted titanium alloys (Table 1).

Properties	Norm (ISO Standars, 2010)	Ti64 (EBM)	Ti64 (cast) (ASTM, 2010)
Yield strenght [Mpa]	760	849	825
Elongation [%]	10	15	10
Area reduction [%]	-	37	15-25
Young modulus [GPa]	-	125	-

Table 1. Comparative view of Ti64 properties: Electron Beam Melting (EBM) vs wrought Ti64.

However, the important advantage of Additive Manufacturing is the capability of producing designed porous material that can be combined with solid material in a single implant. Hence, it is interesting to take a look at the properties of porous material made on EBM. According to a several studies (Petrovic, 2011), (Parthasarthy, 2010), the properties of porous Ti64 fabricated on EBM are comparable to commercially available materials (titanium foam, tantalum, etc.), even approaching to human bone properties as shown in Table 2. [1] [2] [3] [4]

Property	Porous Ti [EBM] [1]	Alternative foams
Porosity [wt %]	57.5	62.5 [2]
Compressive modulus [MPa]	2927	3000 [3]
Compressive strength [MPa]	195	65 [3]
Flexural strength [MPa]	101.98	105 [2]
Tensile strength [MPa]	~78 [4]	70 [2]
Fatigue properties	Fm = 3820 N R = 0.1 N ≥ 5.000.000 cycles	-

Table 2. Comparative view of porous Ti64 properties: Electron Beam Melting (EBM) vs commercially available materials

4.2 CoCr alloy

Cobalt Chromium is commonly used in fabrication of implants that are submitted to intensive wearing (knees, shoulders, elbows, etc.). Hence, it is very important for the material processed on AM to show good wearing resistance. For EBM CoCr that corresponds to ASTM F75 is commercially available. Experiments and tests have been made with this material (Petrovic, 2010) and confirm that the main mechanical properties comply with the corresponding norm (Table 3).

Property H/V	CoCr [EBM]	ASTM F75-07 (ASTM, 2010)
Tensile strength [MPa]	1171/1188	450
Yield strength [MPa]	776/769	655
Elongation [%]	5/7	8
Area reduction [%]	6/8	8
Thermal expansion coeficient [x10^{-6} 1/°C]	14-18/13-17	-
Wearing rate [x10^{-8} mg/cycle]	3.3/3.44	-

Table 3. Comparative view of CoCr F75 properties: Electron Beam Melting (EBM) vs corresponding norm.

5. Biological testing and validation

In addition to good mechanical behaviour, material produced by EBM has good biological response as well. Thomsen et al (Thomsen, 2009) have performed a study of surface characterization and early response of porous EBM material in rabbits. According to this

[1] Data for EBM porous Ti64 with pore size of 504 µm

[2] Data for Ti foam

[3] Data for tantalum foam

[4] Estimated value upon the results of samples with smaller and bigger pore size

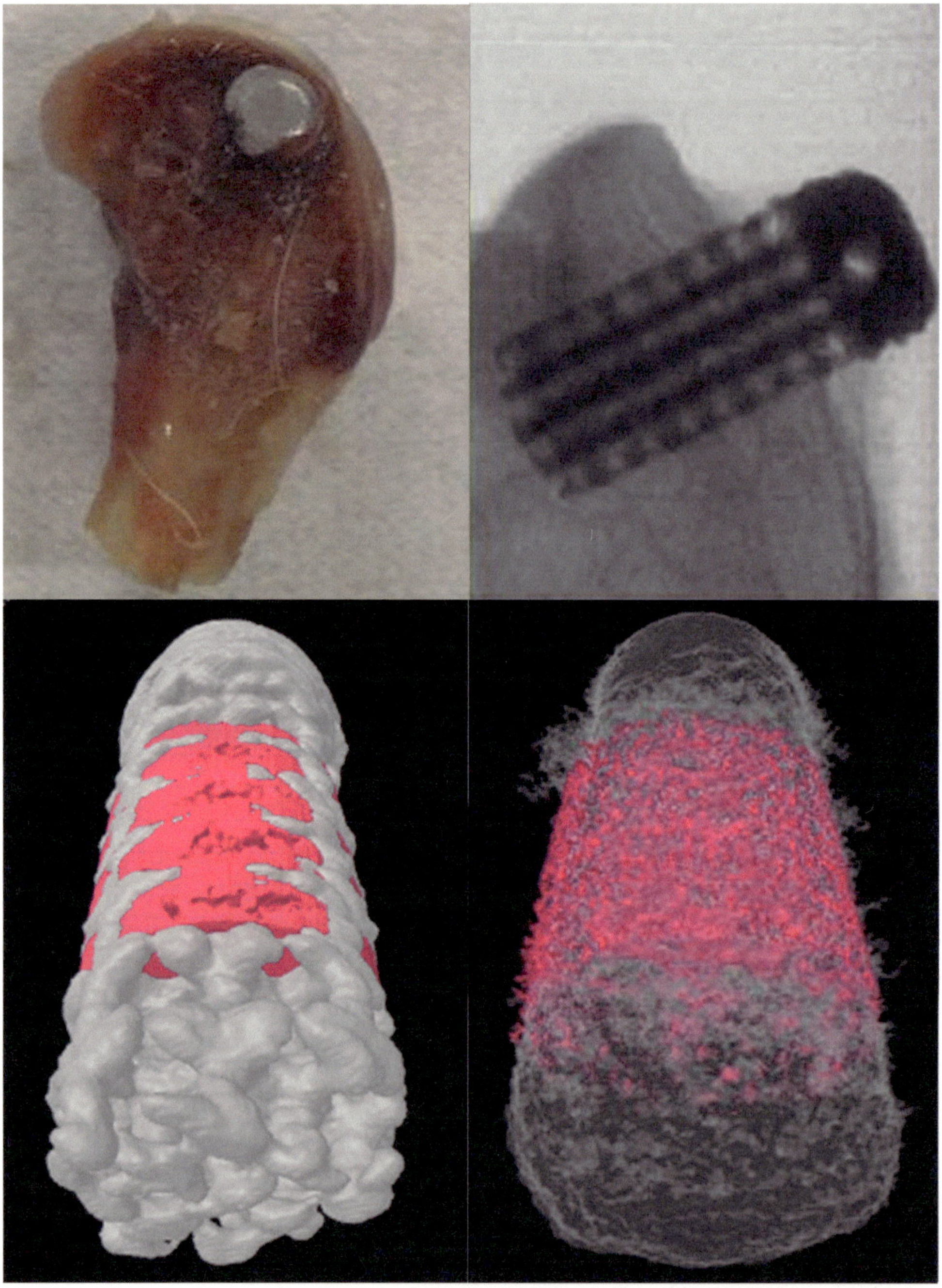

Fig. 14. Results of bone ingrowth testing of porous Ti64 in New Zealand rabbits (Courtesy of Instituto de Biomecánica de Valencia): a) excised sample for pull-out test; b) CT image of EBM sample; micro CT reconstruction of EBM (c) and conventional (d) Ti64 sample.

study the as-produced EBM Ti6Al4V implants had increased surface roughness but similar surface chemical composition compared with machined, wrought Ti6Al4V implants. Also, the general tissue response was similar with a high degree of bone-to-implant contact for all implant types. The results show that the surface properties of EBM Ti6Al4V display biological short-term behavior in bone equal to that of conventional wrought titanium alloy.

Furthermore, the bone ingrowth of EBM scaffolds in rabbits has been evaluated as well (Petrovic, 2011). For evaluation of bone ingrowth, EBM samples were compared to the samples provided by two medical device manufacturers, BIO-VAC and Eckermann. Five samples of each type were implanted in the femur of rabbits (Figure 14b). The control period was 8 weeks. The results show that after 8 weeks between 64 and 86% of void space was filled by the bone tissue. In addition, no adverse effect (infection, inflammation, rejection, etc.) was noticed in animals submitted to this study.

6. Applications

As explained in Supply Chain section, AM technologies manufacture directly from digital information of the part (digital files with 3D geometry) and do not need any kind of auxiliary tooling during the manufacturing process. Normally, the use of tooling (moulds, machining tools, etc) makes crucial influence on the product geometry, since desirable product features cannot be produced.

These manufacturing constraints are not present in AM processes. Using AM processes, designers are not limited or conditioned by conventional manufacturing constraints and can focus only on the optimum design of the product according to its application. AM technologies permit greater freedom in product design, enabling the manufacturing of much more complex geometries and in many cases, geometries that are impossible to manufacture with another fabrication method.

As a matter of fact, EBM has few manufacturing constraints, in terms of producing complex geometries and scaffolds structures and also offers the highest production speed (AIMME, 2009). Its high productivity makes economically viable the fabrication of high added value implants. Although it must be said that due to high processing temperature in EBM process, unfused powder is sintered around the part or scaffold. In certain geometrical features the cleaning process may be difficult, especially in large scaffolds with very small pore size.

The main advantages of using EBM as a Manufacturing technology for implants consist in:

- Full customization. Implant geometry can be customized to the anatomy of the patient and to its specific injury or pathology and fabricated by EBM, with its inherent benefits.
- Controlled and designed porosity. The inclusion of porous regions on the surface of the implant improves the bone osteo-integration in the patient body.

For the time being, three titanium (Ti64, Ti64 ELI and Ti grade 2) and one cobalt chromium (CoCr ASTM F75) alloys are being commercialized by the EBM technology provider and widely used for medical implants. There is a big number of case studies of customized implants that have been implanted in human body. There are also standard implants with added value certified for sale in EU and worldwide. In this section, the authors demonstrate the advantages of AM through different types of application, such as:

- Customized implants.
- Scaffolds with controlled designed porosity.
- Production of small-medium series of value-added biomedical products.
- Production of standard value-added biomedical products.
- Research in the biomedical field.

6.1 Customized implant

Currently, CT imaging has improved, in terms of resolution and 3D details, obtaining very accurate information from the patient. With this information, implants can be designed taking into account patient's anatomy, type of injury, surgical technique, etc. As previously mentioned, the design process starts from scanned information of the patient (Computed Axial Tomography (CAT), Magnetic resonance imaging (MRI) or Radiography). The implant design and development process are also supported by the previous experience of the design team (scientists, engineers, surgeons, etc). The new customized implant design is commonly validated by Finite Elements Analysis (FEA or FEM). In the case of a structural analysis, the solution shows a 3D map distribution with the stress level (strain, displacement, etc) along the geometry (Figure 15).

Principal benefits of customized implants are:

- Implant design adapted to the patient's anatomy and pathology.
- Diminishing of the stress shielding effect.
- Avoiding of manual adjustment of standard implant during surgery.

6.1.1 Design adapted to the patient

Customization of implant geometry is especially important in long duration prostheses (between 10-15 years), i.e. hip and knee prostheses, since the implant geometry adapts perfectly to the anatomy and injury of the patient (Figure 15). In other words, the size and weight of the prosthesis should be the strictly necessary for every patient, increasing the level of comfort. These benefits can be better understood comparing with standard prostheses and usual surgical procedure. When a standard prosthesis is implanted, the surgeon must decide which implant size could fit in the patient. It may be necessary to choose the bigger size. In this case, the surgeon has to create sufficient space for putting in and fixing the prosthesis. It implies to cut and remove more bone tissue and the patient has to carry with a bigger prosthesis. The removal of bone could be critical in the case of younger individuals (< 60 years) (Ratner, 2004), where it might be necessary to perform a revision surgery. Revision prostheses are bigger, causing further increase of fitting place and therefore the removal of more bone tissue. In contrast, the use of customized implants implies that the surgeon has just to create the minimum necessary space in order to fit the implant and leaves intact much more bone tissue for future revisions.

6.1.2 Diminishing of stress shielding effect

It is worth mentioning that customized implants could diminish the negative effect of stress shielding. Stress shielding refers to the reduction in bone density (osteopenia) as a result of removal of normal stress from the bone by an implant (for instance, the femoral component of a hip prosthesis). According to Wolff's law (Wolff, 1986), bone in a healthy person or animal will remodel in response to the loads it is placed under. Therefore, if the bone load decreases, bone will become less dense and weaker because there is no stimulus for continued remodelling that is required to maintain bone mass. During the design process of the customized implant stress shielding can be taken into account and minimized by means of different designs and MEF structural analyses in order to achieve two aims:

- The implant should transfer bearing loads to the bone in order to avoid bone resorption.
- Decrease the stiffness (Young modulus) of the implant in order to make it similar to bone stiffness.

In case of standard implants, the size chosen by the surgeon doesn't adapt perfectly to the biomechanics of the patient and there will be a higher probability of bone resorption since this standard implant doesn't transfer loads properly to the nearby bone tissue.

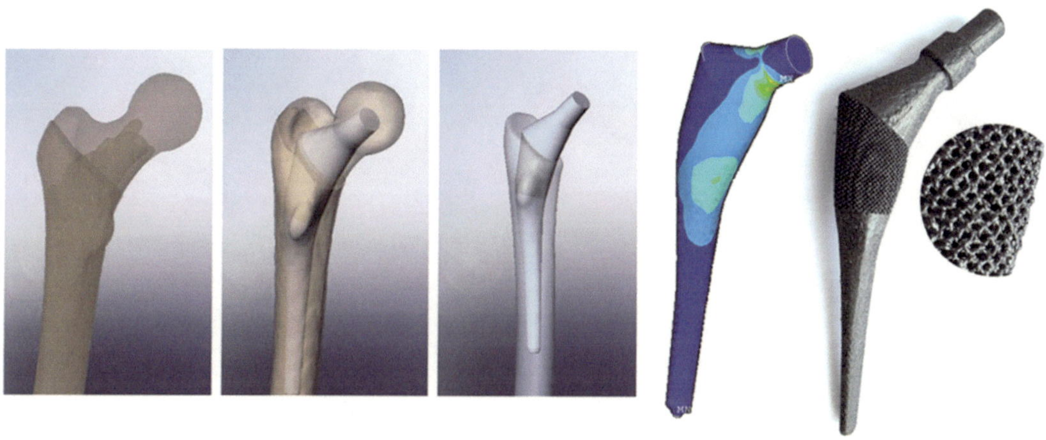

Fig. 15. Customized hip prosthesis developed in Project FABIO (Delgado, 2010). This design was manufactured by means of EBM including a porous region for improving osseointegration and implant fixation in the body. (Courtesy of AIMME, IBV, ASCAMM, TECNALIA).

In conclusion, the use of standard implants not only implies to remove more bone tissue than with tailor-made implants, moreover there is a also higher probability of bone resorption. Therefore, the future revision surgery might be more complex since the patient has lost more bone tissue in the damaged area.

6.1.3 Avoiding manual adjustment of standard implants

Another very important benefit of using customized implants is that surgical operations are shorter. During the operation, surgeon must adjust standard implants manually to the patient's anatomy and pathology in order to be able to place and fix them. This cannot be done until the surgeon has intervened the damaged zone (Figure 16).

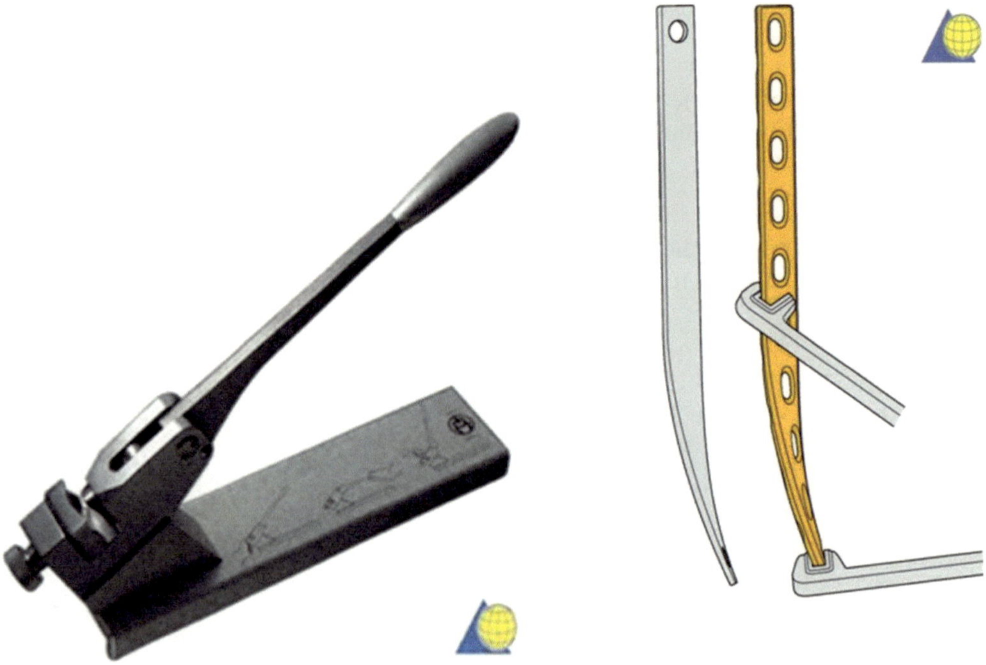

Fig. 16. Tooling for standard plate deformation and adaptation. (Courtesy of AO Foundation Engineering).

In the case of customized implants is not necessary for surgeon to perform these modifications because the implant has been designed and manufactured totally adapted to the patient from direct scanned information (CAT, RMI, etc). It has been proved that the shorter the surgical operation the faster the patient's recovery and this improves patient's life quality. Further advantages of shorter surgical operation are:

- *Lower exposition to external bacteriological agents* (even in sterile atmosphere) leads to lower infection probability (one of major risks during traumatologic and orthopaedic operations).
- *Lower dosis of anesthesia is necessary.* This factor may be critical in certain cases.
- *Less recovery time in the hospital.* Therefore the costs of the surgical operation are reduced and more patients can be treated.

In addition, the adaptation of the conventional implant shape during the surgical operation produces 2 important drawbacks:

- When deformed, the implant has been damaged since the material has suffered plastic deformation. This deformation produces local strengthening while reducing ductility, which leads to the change of implant behaviour. This is especially critical in long term prostheses that bear cyclic loads, since this plastic deformation may reduce its fatigue strength and total implant durability, the implant has suffered a very harmful deformation equivalent to lots of cycles in normal conditions of use.
- The second inconvenience is that corrosion may appear due to the fact that plastic deformation may break the passive layer of the contact area (Pitting Corrosion). Localized corrosion can occur as a result of imperfections in the oxide layer. This can result in a large degree of localized damage because the small areas of active corrosion become the anode and the entire remaining surface becomes the cathode (Ratner, 2004).

After above mentioned, the benefits of using customized implants seem to be clear. Anyhow, the customization design process of an implant can drive to very complex geometries difficult or even impossible to manufacture by conventional manufacturing techniques such as machining, casting, forging, etc. Conventional manufacturing processes can be normally discarded for the following reasons:

- Casting, forging and HPDC are not economically viable due to the investment in the mould only for a unique part.
- In case that machining of customized geometry is possible, the more complex the geometries to produce, the more machining operations and setups are needed and the less economically competitive, comparing with AM.

6.1.4 Case studies of customized plates

In order to show above explained benefits of implant customization and fabrication with additive technologies, several case studies have been made with customized implants for animals. In these case studies, implants have been manufactured using EBM technology and implanted in dogs in a close collaboration of Metal-Processing Technology Institute AIMME, CIMA Research Group (University of Vigo) and FAUNA Veterinary Clinic (Rodiño, 2010), (CIMA, 2011). Customized implants for ten different patients have been developed, three of which are going to be presented in this section. Specific osteosynthesis plates were designed by researchers from CIMA together with the veterinary surgeon using the procedure mentioned in the Supply Chain section - upon medical images of each patient the model was reconstructed and manufactured on the EBM technology to be implanted. AIMME has offered technical advice related with the manufacturing technology (EBM) during the design process, has manufactured implants by means of EBM and also has performed finishing operations in order to get the optimum surface roughness in every case.

Case of BABY

The first patient Baby, is a 3-year old Yorkshire Terrier. He was admitted in December 2009 presenting a radius-cubital diaphysal oblicuous fracture with severe loss of radiographic density due to osteoporosis and disuse of the leg in question. If the X-ray images of healthy and injured leg are compared (Figure 17), it is evident that a 20-25% of radius-cubital length had been lost. This was a challenging experimental case with few chances of successful outcome. Bearing in mind that there wasn't any commercially available plate, of this size and capable of being adapted to the damaged zone, the surgeon and CIMA researchers decided to design a customized titanium plate.

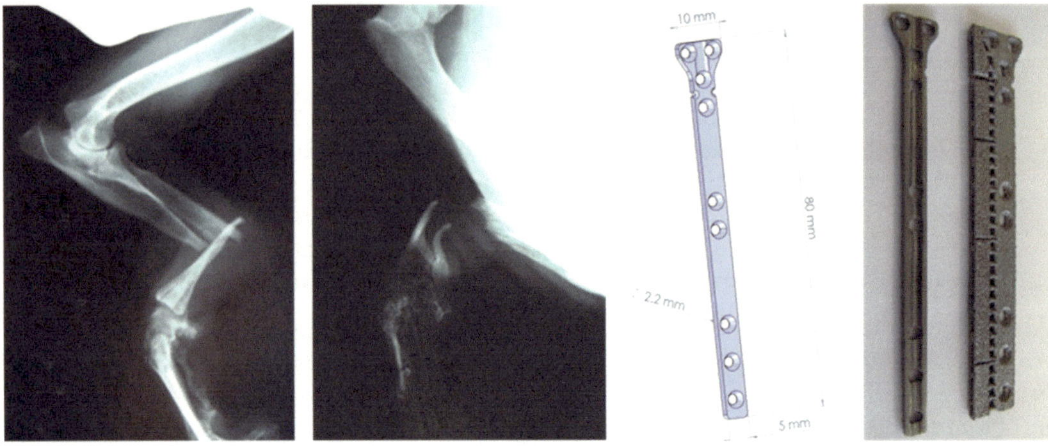

Fig. 17. From left to right: X-ray of the fracture suffered by Baby; customized CAD design of the plate and the plate as built on EBM and finished. (Courtesy of CIMA, University of Vigo and Veterinary Clinic FAUNA www.clinicafauna.es).

Figure 18 shows the outcome of the surgical intervention which took place in February 2010. The surgeon decided to apply an arthrodesis with the aim of preserving the leg and recover its functionality. Fifteen days after the intervention, Baby was submitted to a control of mobility and use of the leg, an almost normal mobility was observed. In addition, Baby's owner reported that, the dog recovered its normal impulses to jump and run, even after much time without being able to walk.

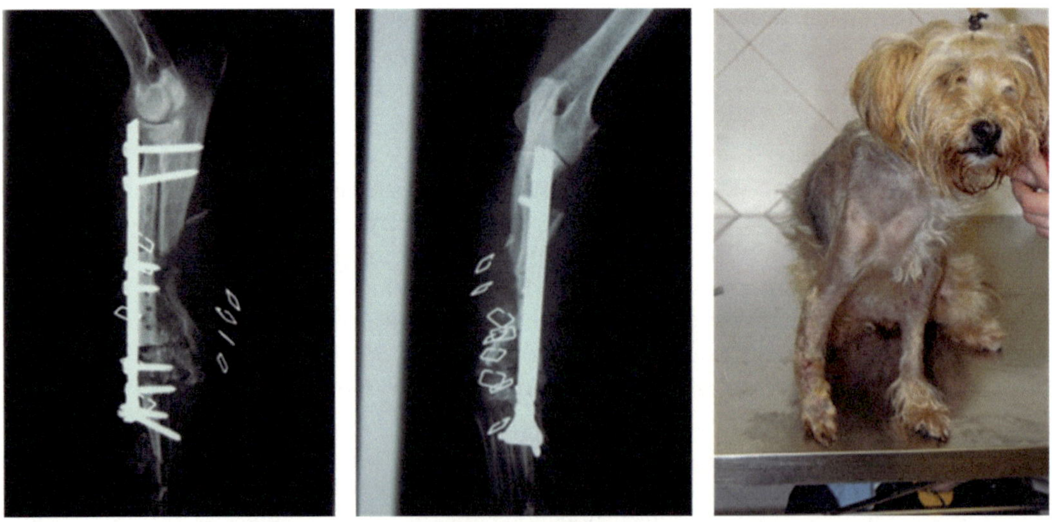

Fig. 18. From left to right: X-rays of Baby after the plate has been implanted and Baby leaning on its injured leg (Courtesy of CIMA and FAUNA).

Case of ARGOS

Argos is an 8-month Bordeaux Mastiff of 50kg. He was diagnosed a bilateral hip dysplasia with grade I patella luxation in the right knee. Initially, the surgeon performed a triple pelvis osteotomy using a standard commercial plate. Only 3 days after the surgical intervention, during the first control, a loosening of fixation screws was detected. The proposed measure was to increase the size of the screws from M3.5 to M4.5 (Figure 19).

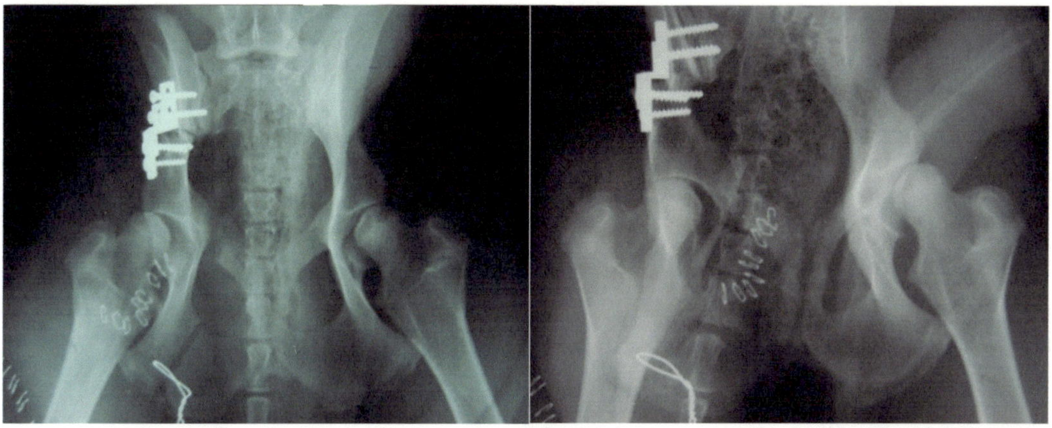

Fig. 19. X-rays after implanting the commercial plate: loose screws after the first intervention (left) and new bigger screws after the second intervention (right) (Courtesy of CIMA and FAUNA).

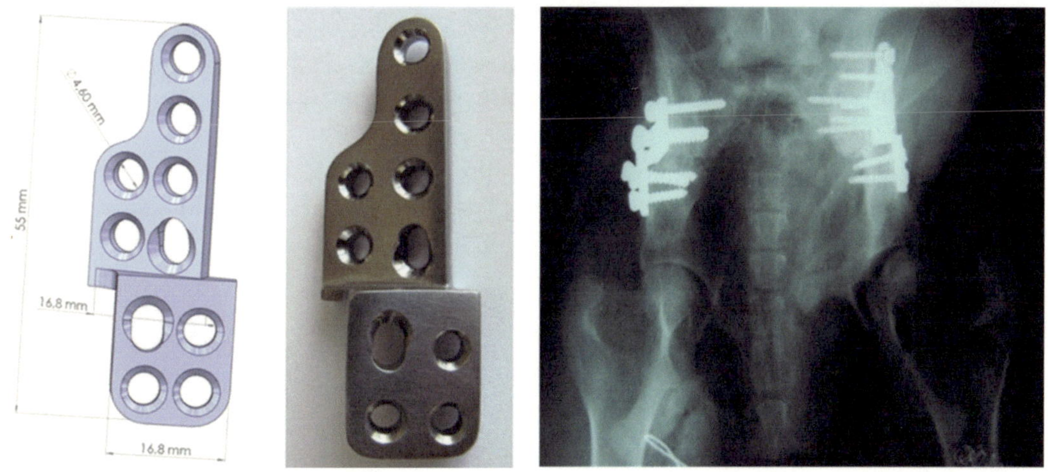

Fig. 20. From left to right: new customized design; plate after being built and finished and X-rays after the intervention (Courtesy of CIMA and FAUNA).

In order to prevent risks of screw loosening, the surgeon sought for a more feasible solution for the left hip side. Due to previous experience with customized plates, the surgeon treated left hip dysplasia with a customized plate made on EBM (Figure 20). As expected, the first revision revealed that the screws on the right hip (commercial plate) started to migrate again while the screws on the left hip (customized plate) stood as tight as after the intervention. It was also observed that a primary consolidation (the best possible) took place thanks to the proper load distribution.

Case of FITO

Fito is a 4-year old German Sheppard of around 40kg. Fito was hit by a car and suffered femur fracture. X-rays showed a multiple diaphysal fracture with big bone splinters. The bone fracture was repaired with a standard titanium plate and 12 screws of M3.5. During the intervention, the surgeon noticed that the plate had to be modified by hand to fit the damaged zone. Surgeon was modelling the plate with a pair of surgical pliers, and verifying visually that the plate was fitting properly. However, as mentioned before, this approach has two principal drawbacks: the surgical intervention is prolonged and the material of the plate suffers plastic deformation, which reduces significantly its fatigue behaviour and even causing precocious plate fracture.

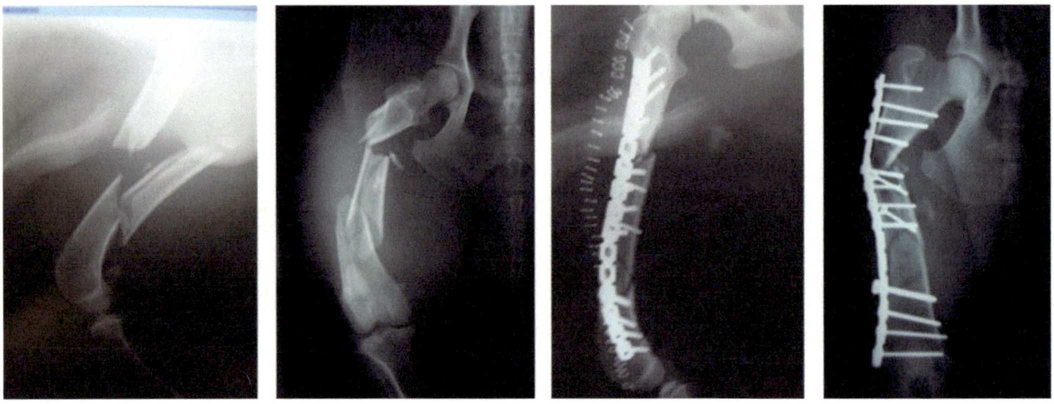

Fig. 21. From left to right: lateral and frontal view of the fracture; X-rays immediately after the intervention and X-rays after the plate fracture was detected (Courtesy of CIMA and FAUNA).

The later drawback appeared after only 3 days, Fito started to limp severely, presenting an angular deformation in the femoral zone. The plate suffered a fracture and completely lost its functionality (as shown on the last image of Figure 21).

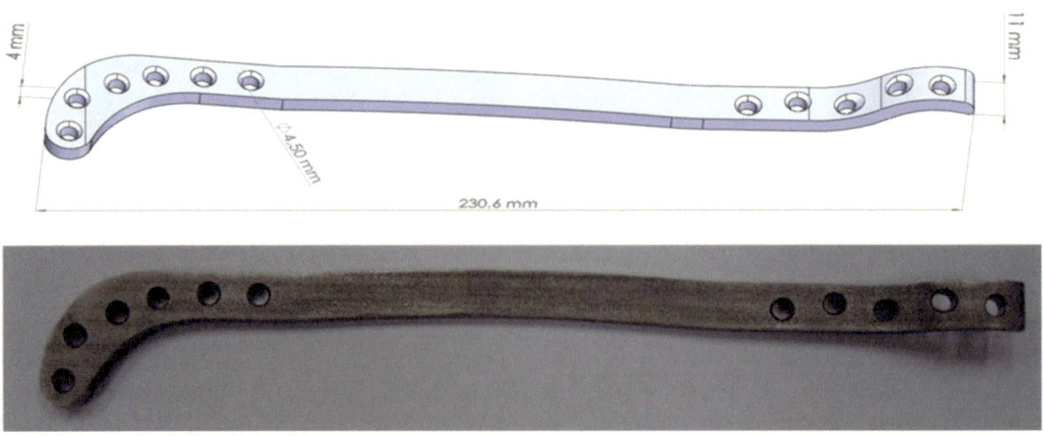

Fig. 22. Customized CAD design (above) fracture fixation plate build on EBM (below).

In order to avoid this problem a specific customized plate was designed and made in Ti64 ELI (Grade 23) ASTM F136 (ASTM, 2010) on EBM (Figure 22).

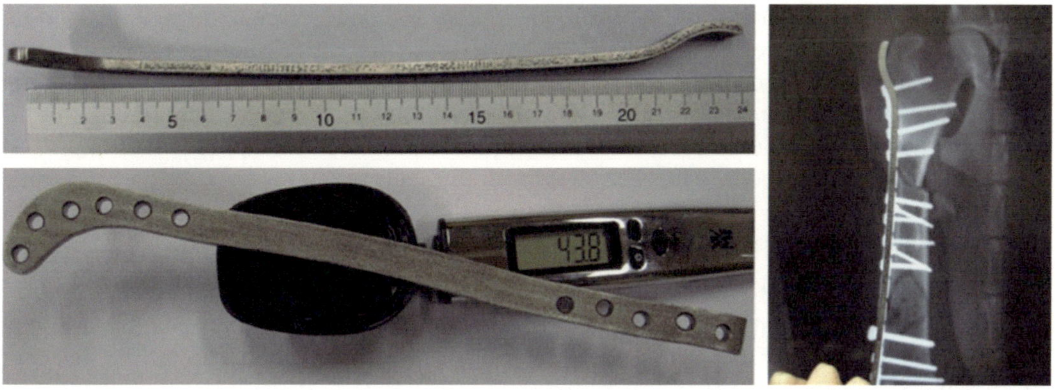

Fig. 23. Left: images showing size and weight of the customized plate. Right: post-surgery X-rays image of the plate (Courtesy of CIMA and FAUNA).

All case studies exposed above have shown that the use of customized implants makes the recovery of patients faster with fewer problems related to posterior revision surgeries, such as screw loosening, plate damage and fracture. In addition, it is important to highlight that, in some of these cases, plates were successfully removed after the bone was recovered, detecting that the bone had completely recovered its functionality.

Regarding Additive Manufacturing, these case studies show that EBM technology is a powerful tool for the manufacturing of customized implants due to:

- its production speed (less than 10 hours in the fabrication by EBM for each customized plate);
- excellent mechanical properties obtained in Ti6Al4V ELI;
- rough surface finish - unlike other surfaces that were polished, the plate surface in contact with the bone was left rough upon surgeon's request to allow better bone fixation.

Accordingly, the EBM process has been validated as one of the main options in case of customized implants: not expecting to replace existing production processes but to be an excellent alternative with certain advantages.

6.2 Scaffolds with controlled designed porosity

As exposed previously, EBM technology enables one-step manufacturing of prostheses that combine solid and porous zones (scaffolds). By means of 3D CAD software, these scaffolds can be designed with the desired pore size, morphology, well-interconnected porosity and gradual transition from solid (body implant) to porous (scaffold), as shown on Figure 24. Designers can control the implant design and have freedom in the design of scaffolds (multiple geometrical solutions) for different pathologies. Only AM technologies are able to manufacture this kind of 3D geometries. For the time being, implants are coated with additional post-processes as plasma spray, microspheres sintering, etc. In contrast with EBM technology, these coatings are not able to produce regular or controlled scaffolds and provide less freedom in design (Ratner, 2004).

In bimetallic implants, there is a substantial risk of galvanic corrosion generated when materials with different electronegativity are placed in the same solution (Ratner, 2004), (Pedeferri, 2007). Conventionally made implants are usually coated with different methods and materials (plasma-sprayed titanium, Ti wire mesh, porous coatings made of CoCr or Ti, etc) for creating rough surfaces for encouraging bone ingrowth and proper fixation. With these conventional methods different metals might be combined and risk of galvanic corrosion may appear. In EBM, there isn't such a risk, due to the fact that part and scaffold are manufactured in the same material in one-step process (Figures 1 and 24).

Porous regions in contact with the bone tissue promote osseointegration (direct structural and functional connection between living bone and the surface of a load-bearing artificial implant), creating better fixation between the prosthesis and the bone. Due to the fact that scaffolds provide void space for bone ingrowth (Figure 14), this bone ingrowth will enhance the fixation of the implant. Scaffolds also contribute to transferring loads between the implant and the bone, avoiding the previously mentioned effect of stress shielding.

Fig. 24. Porous cranial implant (Courtesy of Arcam), customized acetabular cup (Courtesy of Arcam) and hip stem (FABIO Project) (Delgado, 2010).

EBM technology enables manufacturing of implants with different kind of scaffolds. These scaffolds can be placed in different regions of the same implant and each scaffold could have different features (pore size, morphology, etc) if the application requires. In addition, it is possible to manufacture totally porous implant (Figure 24).

Summarizing, EBM enables the manufacturing of implants with scaffolds, designed by means of CAD 3D tools (controlled porosity), which enhances the development of a new typology of high added value implants that are designed and manufactured according to the needs of the patient and not the other way around. This is possible because the EBM process doesn't have constraints imposed by traditional manufacturing processes. So as to illustrate above mentioned discussion, in the next paragraphs several kind of applications and high added value implants are going to be presented.

6.3 Production of standard value-added biomedical products

EBM technology is being used by biomedical industry for production of large series of standard implants (different sizes) but with the added value of a 3D scaffold designed by computer. These are successful cases of Italian manufacturers Ala Ortho and Lima, with different products in the market. They manufacture acetabular cups with different 3D scaffolds which offer better response in the human body. These products have obtained the CE Mark and are being implanted in humans (Figure 25).

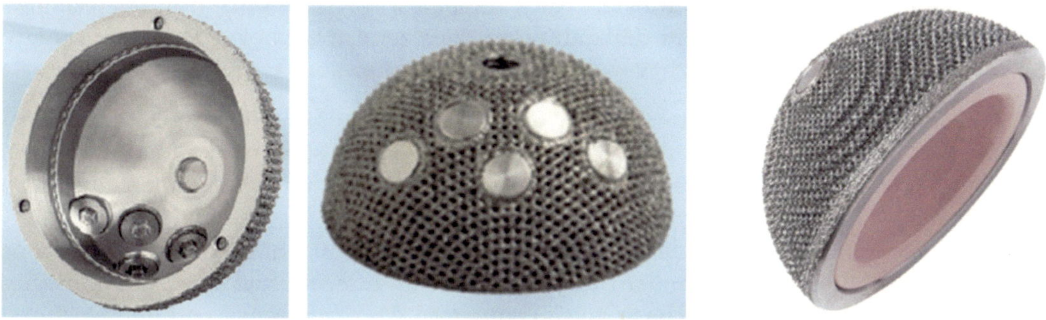

Fig. 25. Acetabular cup implants made on EBM by Italian companies: Cotile Fixa Ti-Por® by Adler Ortho® (left & middle) and DELTA-TT® Cup by Limacorporate (right) are already available on the market.

6.4 Production of small/medium series of value-added biomedical products

EBM technology can also be used for the production of small/medium series of implants with a competitive price, since it offers an interesting alternative to other manufacturing processes where previous investment in tooling is necessary. A variety of industrial sectors use tooling for manufacturing of large series (millions of units) where the cost of the tooling is included in the final price of each produced part. The larger the production, the smaller the cost fraction included in each part. This is not economically viable in the case of small-medium series, since the cost fraction derived from tooling increases as the number of parts decreases. The case of a customized implant is the low end of small series production (Figure 26).

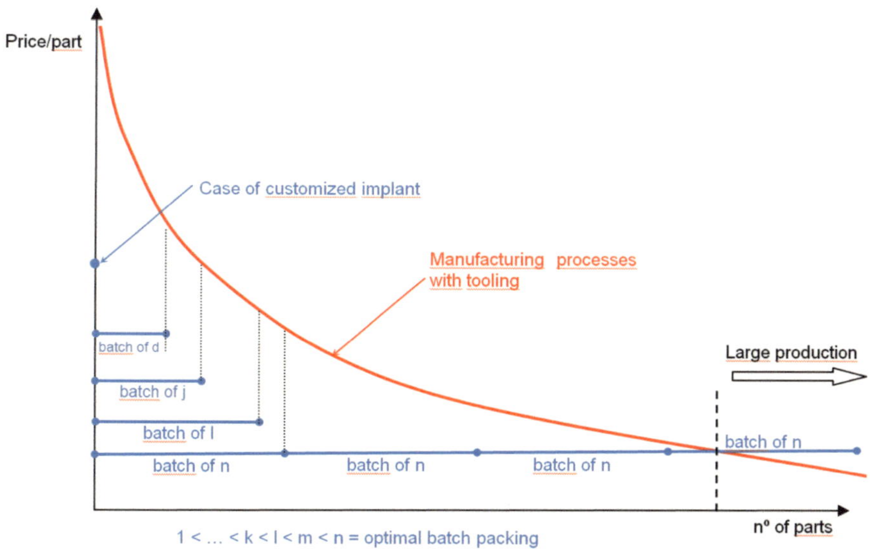

Fig. 26. Comparison of the cost per part using manufacturing processes with tooling and AM technologies.

In the case of manufacturing implants with EBM technology, there is no need of tooling, since EBM uses the digital file of the implant (3D CAD) for its direct fabrication as described in section 2. As a consequence, EBM makes profitable the production of small/medium series production because the cost of the batch remains always the same (for a determined geometry and batch size) without depending on the size of the whole production (Figure 26). All these economical aspects have several positive implications for EBM:

- For manufacturing of greater variety of implant sizes on EBM, unlike traditional manufacturing processes, it is not necessary to have a huge stock of moulds and tooling, since 3D models of different sizes are stored electronically and manufactured on demand.
- Semi-customization of standard products is also viable on EBM. Small modifications can be introduced in the standard implant model in order to treat specific pathologies with special requirements. The Spanish company LAFITT designed and manufactured a Total Hip Prosthesis with special indications (ASTM, 2010). Ten prototypes were manufactured by EBM in Ti6Al4V ELI in 36 hours. These prototypes are being evaluated by several tests. If results from tests are positive LAFITT may consider EBM technology as a production method for medium-small series, case of prostheses with special indications (Figure 27).
- The gradual evolution of the implant design and the addition of improvements between a fabrication and the next one are possible, due to the feedback and experience from already implanted cases, new findings, competitors, etc. At this point implants manufacturers that use conventional manufacturing techniques have two options:
    - Make costly tooling corrections in order to improve some features and keep product competitive and up-to-date or
    - Keep producing parts with the original implant design until the mould is paid off by a determined number of sold units. Then, the design of the implant can be updated by corrections on the mould. The fact that for standard implants there are several sizes implies one mould for each size. More demanded sizes can be updated earlier than the others because of moulds investment depreciation.
- As mentioned before, AM favours manufacturing batches in which every part is different in size, shape, special features, etc. This is possible because lots of parts can be packed inside EBM process chamber which has a capacity of 200x200x350 mm. One example is the case developed by the Spanish company SURGIVAL. This company has designed osteosynthesis plates for the treatment of different fractures in the human body (Radius distal, Proximal and Proximal Humerus plates). In total, 21 plates (different models and sizes) were manufactured in 7.5 hours by EBM in Ti6Al4V ELI fulfilling the standard ASTM F136 (grade 23). These plates are shown on Figures 28 and 29.

6.5 Rapid prototyping for the research in the biomedical field

EBM is being widely used for Research & Development in the biomedical field due to its production speed and design freedom which permits shortening the development periods. For this reason it is used for manufacturing prototypes in order to validate new products and concepts. As an example, the Spanish company SURGIVAL has developed a hollow tibial prosthesis for knee arthroplasty. The prosthesis has sensors for detecting osseous loosening by means of telemetry. For this application this tibial prosthesis was manufactured in 6 hours by EBM in Ti6Al4V, consisting of two parts: the main body and a cover for the introduction of sensors inside (Figure 30).

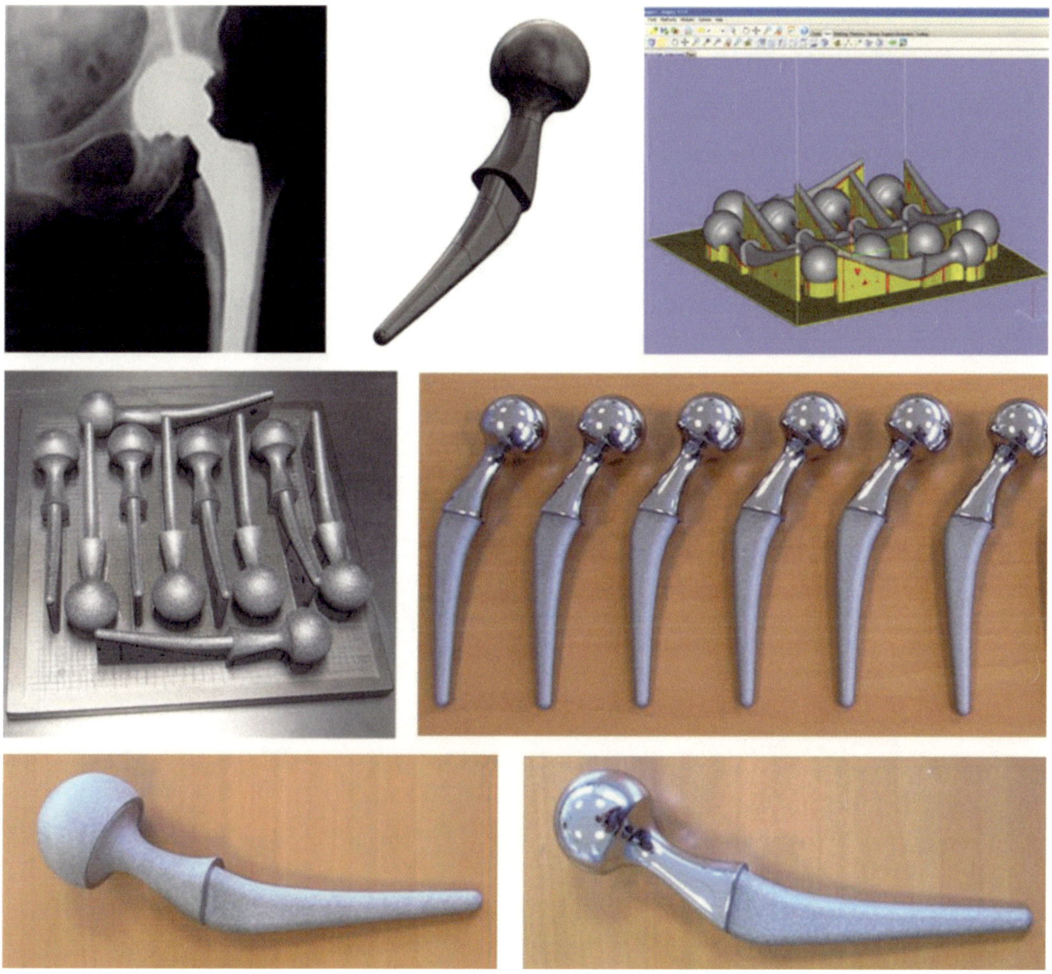

Fig. 27. LAFITT case. Upper row: radiography of a Hip prosthesis implanted; 3D image of the design of a Total Hip prosthesis for special indications; Virtual preparation of batch for building in Magics®; Middle row: 10 prototypes manufactured by EBM and complete batch of 10 prostheses finished ready for testing; Downer row: Comparison between Hip prosthesis before and after polishing.

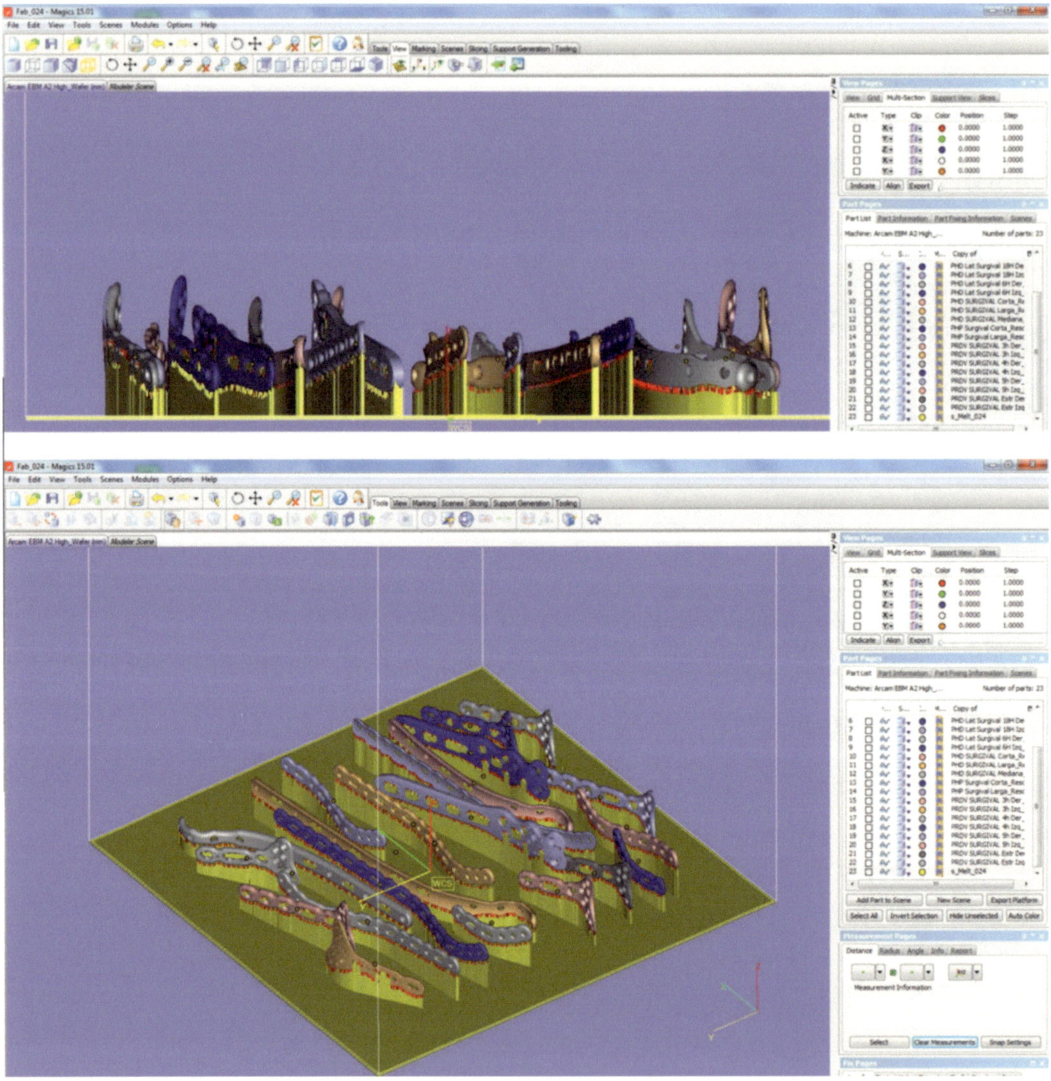

Fig. 28. Virtual preparation of batch for EBM building: case of 21 customized radius distal, proximal and proximal humerus plates manufactured in a single build (Courtesy of SURGIVAL).

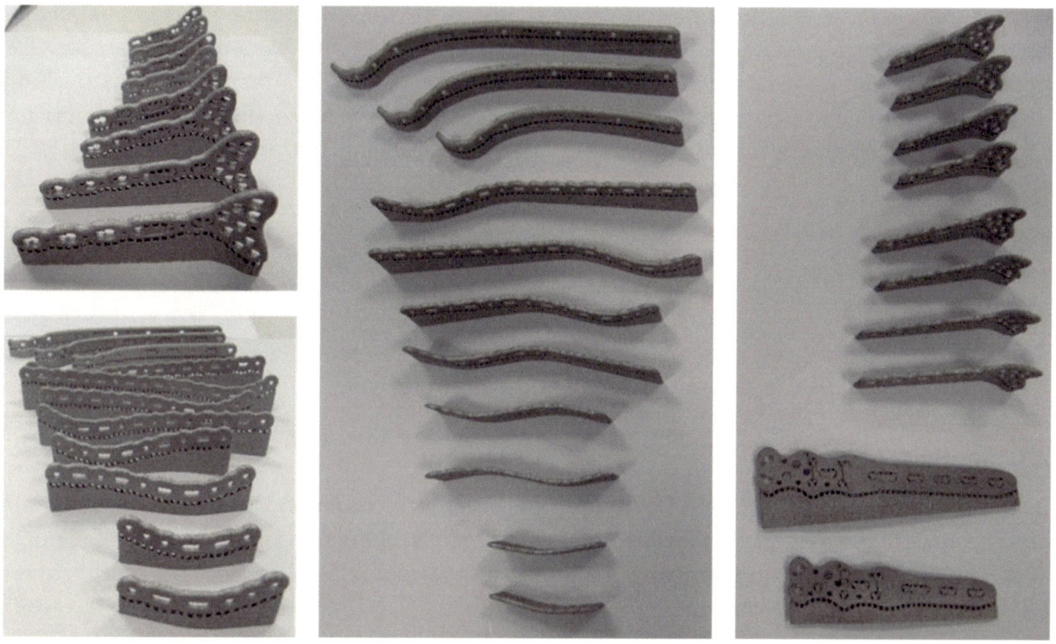

Fig. 29. SURGIVAL case. 21 customized radius distal, proximal and proximal humerus plates manufactured in a single EBM build (Courtesy of SURGIVAL).

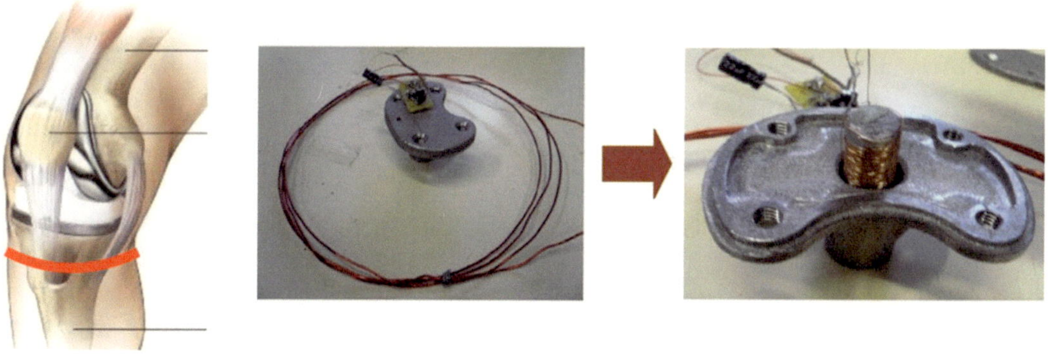

Fig. 30. Sensorized Tibial prosthesis manufactured by EBM in Ti64 for detection of osseous loosening. INTELIMPLANT CDTI's Project (Courtesy of SURGIVAL).

7. Future developments

As shown in the Applications section, EBM technology has proved to be a very powerful tool for manufacturing of high added value products. At the same time, EBM is a relatively recent manufacturing technology (first machine sold in 2002) and has much more room for improvement. For both reasons, there are different R&D attempts being performed along the world in order to obtain new findings applicable to new biomedical implants and for supporting the biomedical industry. These attempts are being performed in different R&D areas, such as:

- Materials
- EBM Technology
- Software

The following advances are not relevant only for biomedical sector but do offer competitiveness to additive manufacturing of medical implants.

7.1 Materials

In the biomedical field, lots of efforts are being made in order to develop new materials and/or improve material properties (metals and polymers) for biomedical applications. Also, the evolution of manufacturing processes, Coatings and Surface Implant Modification are remarkable areas of work.

Nowadays, metals are the best option for long time load-bearing prostheses. For this reason, the development of new metallic biomaterials for long-term prostheses and adequate the manufacturing processes in order to achieve best possible properties is a matter of great interest.

In particular, Titanium alloys are very interesting for research due to its excellent properties such as high corrosion resistance, low modulus, high fatigue strength, low density (lower than most common metallic materials), good mechanical properties, etc. There are two relevant areas under development:

- *Biocompatibility improvement:* Some attempts are being done for improving traditional Ti6Al4V alloys, i.e. Ti6Al7Nb and Ti5Al2.5 alloys improve material biocompatibility and mechanical properties, substituting Vanadium by Nb or Fe respectively (less toxic). These alloys have similar mechanical properties to traditional Ti6Al4V but providing with higher biocompatibility and slightly lower modulus. (Ratner, 2004), (ASTM, 2010), (ASTM, 2010).
- *Lower Modulus.* Beta-type Ti alloys with a low elastic modulus has proved to be effective for inhibiting bone absorption and enhancing bone remodelling. The addition of some alloying elements like Mo, Zr, Ta, etc permits to stabilize the BCC (beta) phase at room temperature. Moreover, it is well known that the Young modulus of β-type Ti alloys are considerably smaller than those of the α- and (α + β)-type Ti alloys. (Ratner, 2004), (ASTM, 2010), (Niinomi, 2008).

Although for the time being these alloys are not commercially available for EBM technology there are many efforts to adjust the processing parameters (AIMME, 2011).

7.2 EBM technology

As a proof of the improvement potential of this technology, recently a new manufacturing beam strategy named MultiBeam® has been developed and released. This strategy allows EBM to produce finer details and obtain better surface finish, splitting the high energy electron beam in multiple finer beams (less power/beam) so the energy input in every location on each layer can be accurately controlled. MB strategy opens the possibility of manufacturing implants with better surface finish and finer scaffolds (Figure 12). Some attempts with finer beams and powders are being performed in order to achieve higher resolution with the EBM process.

7.3 Software developments

Great efforts in different areas are being made in software developments since AM technologies work with digital files of parts.

- New 3D CAD (Computer Aided Design) tools will permit more freedom in design, since most commonly used commercial 3D CAD tools are conceived for conventional manufacturing processes and don't allow to design easily new concepts with complex geometries, i.e. scaffolds, fractals or bionic features. There are several commercial software tools available, i.e. 3-Matic® which allows common Computer Aided Design (CAD) operations on 3D anatomical data (STL format), Magics®, Netfabb® or AutoFab® which permit automated design of scaffolds structures. Further improvements are being developed in these commercial tools.
- KBE tools. Expert systems for process planning automation which will permit the automated orientation and location of a batch of different parts in the building virtual platform in order to obtain desired features or properties on each part, surface finish, mechanical properties, etc. (Petrovic, 2010).
- Specific CAE tools, for scan to part reconstruction (Mimics®), automatic guidance in implant design in STL format (3-Matic®), assisted an automated topological optimization of lattice structures (Within®), etc. Further improvements are being developed in these commercial tools.

8. Acknowledgments

Authors would like to show their sincere gratitude to CIMA research group from University of Vigo, Veterinary Clinic FAUNA from Vigo, medical device manufacturers LAFITT and SURGIVAL from Valencia. Without their collaboration, this manuscript would not be possible.

9. References

AIMME. (2009). MEDIFUTUR Project: Informe de inspección de vertebras - AIMME, Unidad de análisis 3D, AIDO, November 2009.

AIMME. (2009). Puesta en marcha de una línea de investigación para la fabricación de piezas con aleaciones no férricas mediante técnicas de conformado de material por adición. Internal technical report by AIMME, 2008-09 financed by IMPIVA (Institute for Small and Medium Industry of the Generalitat Valenciana, Spain).

AIMME. (2010). GESDIFFF project: Gestión en diseño para Free Form Fabrication. Internal report of a project financed by Spanish Ministery of Industry and Commerce within "Innoempresa Suprarregional" call, 2010.

AIMME. (2011). BIOMETAL, Estudio de aleaciones metalicas para el sector sanitario procesadas mediante fabricación aditiva. Internal technical report by AIMME, 2010-12 financed by IMPIVA (Institute for Small and Medium Industry of the Generalitat Valenciana, Spain).

ASTM. (2010). F1295 - 05, Standard Specification for Wrought Titanium-6Aluminum-Niobium Alloy for Surgical Implant Applications (UNS R56700), copyright ASTM International, 100 Barr Harbor Drive, West Conshohocken, PA 19428, USA.

ASTM. (2010). F136-08, Standard Specification for Wrought Titanium-6Aluminum-4Vanadium ELI (Extra Low Interstitial) Alloy for Surgical Implant Applications (UNS R56401), copyright ASTM International, 100 Barr Harbor Drive, West Conshohocken, PA 19428, USA.

ASTM. (2010). F1472-08, Standard Specification for Wrought Titanium-6Aluminum-4Vanadium Alloy for Surgical Implant Applications (UNS R56400), copyright ASTM International, 100 Barr Harbor Drive, West Conshohocken, PA 19428, USA.

ASTM. (2010). F1713-08, Standard Specification for Wrought Titanium-13Niobium-13Zirconium Alloy for Surgical Implant Applications (UNS R58130), copyright ASTM International, 100 Barr Harbor Drive, West Conshohocken, PA 19428, USA.

ASTM. (2010). F1813-06, Standard Specification for Wrought Titanium–12 Molybdenum–6 Zirconium–2 Iron Alloy for Surgical Implant (UNS R58120), copyright ASTM International, 100 Barr Harbor Drive, West Conshohocken, PA 19428, USA.

ASTM. (2010). F2066-08, Standard Specification for Wrought Titanium-15 Molybdenum Alloy for Surgical Implant Applications (UNS R58150), copyright ASTM International, 100 Barr Harbor Drive, West Conshohocken, PA 19428, USA.

ASTM. (2010). F2792-10 Standard Terminology for Additive Manufacturing Technologies, copyright ASTM International, 100 Barr Harbor Drive, West Conshohocken, PA 19428.

ASTM. (2010). F75-07, Standard Specification for Cobalt-28 Chromium-6 Molybdenum Alloy Castings and Casting Alloy for Surgical Implants (UNS R30075), copyright ASTM International, 100 Barr Harbor Drive, West Conshohocken, PA 19428, USA.

Biomet Orthopedics. (2009). Patent-Matched Implants. Technical brochure, 2009, Biomet Orthopedics, P.O. Box 587, Warsaw, IN 46581-0587.

Bronzino J et al. (2006). The Biomedical Engineering Handbook - Medical Devices and Systems, 3rd edition, CRC Press, Taylor & Francis Group.
CIMA. (2011). "Deseño, modelización parametrizada e fabricación de próteses para traumatoloxía e cirurxía ortopédica para aplicacións no eido veterinario". Internal technical report by FAUNA Veterinary Clinic and CIMA Research Group (Univ. Of Vigo), 2009-12 financed by XUNTA de GALICIA Conselleria de Economia e Industria, Spain.
Delgado J. et al. (2010). FABIO project: Development of innovative customized medical devices through new biomaterials and additive manufacturing technologies. 3rd Int Conference on Additive Technologies, iCAT '10, 22-24 September 2010 Nova Gorica, Slovenia.
ISO Standards. (2010). Ti6Al4V-ISO 5832-3 "Implants for surgery - Metallic materials - Part 3: Wrought titanium 6-aluminium 4-vanadium alloy", ISO Consensus Standard.
Niinomi M.(2008). Metallic biomaterials, J Artif Organs 11:105–110.
Oshida Y. (2007). Bioscience and bioengineering of titanium materials. Elsevier.
Parthasarathy J. et al. (2010). Mechanical evaluation of porous titanium (Ti6Al4V) structures with electron beam melting (EBM), Journal of the mechanical behavior of biomedical materials 3 (2010) 249 - 259.
Pedeferri P. (2007). Corrosione e protezione dei materiali metallici. Editore Polipress.
Petrović V et al. (2010). Knowledge assisted rapid manufacturing (KARMA project). Proceedings of AEPR'11, 16th European Forum on Rapid Prototyping and Manufacturing. Paris, France, 23-24 June 2010.
Petrović V et al. (2011). A study of mechanical and biological behavior of porous Ti6Al4V fabricated on EBM. Innovative Developments in Virtual and Physical Prototyping - Proceedings of VRAP 2011, 28 Sep - 01 Oct, Leiría, Portugal.
Petrović V. et al. (2011). Additive Layered Manufacturing: Industrial applications through case studies. Int. Journal of Production Research 49(4)
Ratner B. (eds.) . (2004). Biomaterials Science - An Introduction to Materials in Medicine, 2nd ed. Elsevier Academic Press, 2004.
Reeves P. (2008). ATKINS: Manufacturing a Low Carbon Footprint: Zero Emission Enterprise Feasability Study, 2nd International Conference on Additive Technologies (iCAT), 17-19th September 2008.
Rodiño Janeiro B. (2010). Design, modelling and manufacturing of osteosynthesis plates for veterinary traumatologic surgery, Master Thesis, ETSII - University of Vigo, Spain, July 2010.
Synthes. (2007)® Instruments and Implants. Cervios and Cervios chronOS. Radiolucent cage system for anterior cervical interbody fusion. Technical guide.
Thomsen P. et al. (2009). Electron beam-melted, free-form fabricated titanium alloy implants: Material surface characterization and early bone response in rabbits, Journal of Biomedical Materials Research Part B: Applied Biomaterials Vol 90B, Iss 1, pg 35–44 (2009).
Wohlers T. (2010). Wohlers Report 2010: Additive Manufacturing State of the Industry - Annual Worldwide Progress Report, Wohlers Associates, Inc., ISBN 0-9754429-6-1.

Wolff J. (1986). The Law of Bone Remodeling. Berlin Heidelberg New York: Springer, 1986 (translation of the German 1892 edition)

Part 4

Public Perception of Biomedicine

8

Crossings on Public Perception of Biomedicine: Spain and the European Indicators

Eulalia Pérez Sedeño and María José Miranda Suárez
CSIC
Spain

1. Introduction

In recent years, there has been an increasing advance in various disciplines such as genetics, biotechnology or information technology. There is almost unanimous agreement that the generation and application of scientific and technological knowledge is playing a key role in the improvement of quality of life in society, productive modernization, and the insertion of some countries on the global stage. However, these rapid changes are also having serious effects, leading to discussions about their current or future use and their social, ethical implications. To date, these changes have also modified our environment.

Universal scientific knowledge and its recent development have been enabling a techno-economic universality. But questions have also been raised about the deepening of social inequalities and the asymmetrical appropriations of knowledge. At the same time, our societies have experienced major political development that has opened up all areas of public policy to social scrutiny and citizen participation. Despite this, science and technology are still perceived as something distant by some citizens. Still, intense activity in the area of scientific communication and popularization in the last decade may be changing that perception. That is why it is necessary to open public policies on science and technology to the sensitivities and opinions of the people who are affected and interested, to facilitate the practical viability of innovation and depth in the democratization of the systems. The studies of public perception of science and technology are taking a leading role in these aspects.

As we saw earlier (Pérez Sedeño and Miranda Suárez, 2008), studies on the public perception of science and technology originate in the Anglo-Saxon world, with the movements *Scientific Literacy* and *Public Understanding of Science*. The first is a North American movement that seeks to measure the degree of scientific literacy in society, designing surveys where basic scientific questions are asked about well-established facts. That is, questions about content are posed, without taking into account the complexity of the scientific activity. But science isn't only knowledge in the sense of 'information' about facts and pieces of information; of extreme importance are procedures, processes and the nature of the knowledge according to the subjects and techniques applied, as well as the social values in which they are expressed.

The second movement mentioned, fundamentally of British origin, seeks to assess the capacity of society to understand science, its applications and its relationships to society, and as such the questions asked aren't of scientific content, but rather of a social, political or economic type. So, the more traditional semantic component of the notion of scientific culture, which is reduced to the level of scientific knowledge, is put into question. This other orientation appears more appropriate, as the notion of scientific culture already includes communicative skills and abilities, which also entail outlining a type of culture relative to the organizational forms of scientific production and, above all, its interactions, which also form part of the processes of the *public perception of science*.

Starting from the 1990s, and owing to some controversies arising around certain technologies, a new movement was added: that of science in society. At times, science appears distant and unconnected from society; its objectives aren't understood. Lack of a common language and the rapid progress experienced in certain areas of investigation has increased public concerns, and has contributed to people viewing, with preoccupation or ambivalence, the role that science and technology play in daily life. But science doesn't work in isolation. It is developed in concrete situations, historical and social, and scientists are subjects that produce situated knowledges. One of the main contributions that favored the growth in objectivity via the democratization of knowledge and scientific practices has been the role of feminist epistemologies, represented by the feminist point of view of Sandra Harding, the contextual empiricism of Helen Longino and the situated knowledges of Donna Haraway. They deny both neutral science practiced by an universal subject, as well as epistemological relativism. And they champion the democratization of science, consistent in the incorporation of a plurality of perspectives, by way of the annulment of personal idiosyncrasies, a process called cognitive democracy. In this, consensus is the result of a critical dialog where all the relevant perspectives are represented and where obtaining knowledge, scientific-technological advances, isn't an objective in its own right. As such, in recent years an interest in public participation and governance has developed, not just in political discourse and academic analysis, but also in scientific and technological development. Controversies like that of Genetically Modified Organisms (GMOs) or important crises like that brought on by Bovine Spongiform Encephalopathy (BSE), the so called Mad Cow Disease, have reduced public confidence in how scientific and technological developments are handled.

The European Union (EU) has made this movement its own to the point of transforming its program *Science, Economy and Society* into *Science in Society*, within the Seventh Framework Programme (FP7). The explicit objective is to "'bring science closer to citizens' (European Commission, 2002: 7), in order to 'provide a space for scrutiny and informed debate on important issues of public concern'"(European Commission, 2002: 17), but also to analyze not only the inherent benefits but also the dangers, limits and failures of science. In this sense, knowing what citizens think of science and technology, of its benefits and detriments, and of scientists, is vital.

One of the best instruments to do this are surveys of the public perception of science and technology. These surveys respond to a dual motivation, theoretical and practical. On the one hand, it's about improving understanding of the scientific/cultural situation as an important aspect of the general culture of a country, a region or a specific collective. On the other, it's also about making relevant information available for practical ends like, for example, to value the potential popular support for measures that increase public spending on R+D or that establish certain priorities for programs on investigation, innovation, etc.

Works done on the public perception of science have taken form thanks to combined and parallel work relating to North American and European surveys done by Jon D. Miller's investigative groups in the United States and by John Durant in Great Britain. Their emphasis on specifying dimensions of concrete analysis in comparable questionnaires helped these investigations to extend throughout Europe and other countries, so that in the 1990s they were already beginning to have a significant level of empirical foundation.

In the last decades, several researchers have been carrying out periodic surveys on interest, perception and public opinion about science and technology in general, or just some particular aspects of them. Within the United States, The National Science Board of the National Science Foundation (NSF) prepares the Science and Engineering Indicators report on a biannual basis. With this, not only have they continued to carry out surveys on public attitudes towards science and technology since the 1970s, but they also consider promotional strategies and recommendations to incorporate into national policies. In the European experience, the role of the European Commission is important in implementing action frameworks through programs like the Forecasting and Assessment of Science and Technology (FAST program). This program sought to predict and analyze the consequences of the incorporation of new technologies in the Framework Programs of R+D. Hence, the emergence of specific analytical lines, such as robotics or biotechnology, in Eurobarometers allowed to measure questions of understanding of science at European level in recent times. The specific choice of public understanding of science as the study of opinion and attitudes from the Eurobarometer from 1992 to 2010 is essentially due to three reasons. Firstly, decisions influenced by science increasingly make up a more direct part of our everyday acts, albeit unconsciously. Moreover, for an advanced society to effectively develop and participate in decisions that affect it, it is essential that a minimum scientific culture extends horizontally across it. Finally, in the current society of knowledge, scientific training of citizens is increasingly a requirement of democracy.

In the Ibero-American area, although they have been conducting studies of understanding for more than twenty years, it is only recently that they began to conduct standardized surveys on a regular basis. In this sense, the Organization of Ibero-American States (OEI, Organización de Estados Iberoamericanos) and the Network of Indicators on Science and Technology (RICYT, Red de Indicadores de Ciencia y Tecnología) have promoted these types of comparative studies, progressively achieving institutional support such as the Spanish Foundation for Science and Technology (FECYT) or Centro REDES (Argentina), among others. These three institutions now have a priority objective, namely, to attain an Ibero-American standard of indicators of social understanding and scientific culture, which is in the development stage.

In Spain, the FECYT has carried out national surveys on public understanding of science and technology biannually since 2002, and as in the Eurobarometers, the topics of biotechnology, biomedicine and health are covered separately. These surveys usually measure three different levels of the public relationship with science: degree of interest and information on issues of science and technology, level of scientific knowledge, and attitudes towards science and technology. Other entities such as the BBVA Foundation have also realized some surveys about specific areas of biotechnology and biomedicine (May, 2008).

A few years ago, a compilation was published that collected contributions from the most representative figures in the study of social perception of science and scientific culture, under the suggestive title of *Between Understanding and Trust. The Public, Science and Technology*. The volume covers a wide range of subjects, conceptual approaches, methodologies and proposals for disciplinary renovation where, despite their diversity, what stands out is a recurrent agreement that is underlined in the conclusion by the editors: analysis of science's credibility and the trust it arouses in citizens should be considered as the most significant points for the future agenda of investigation.

2. Spain and Europe: Generalized optimism?

Next, we will see some results from the Eurobarometers and FECYT surveys from 2006 and 2010. The principal objectives of the Eurobarometer have been to measure the attitudes of European citizens regarding collaborative investigative projects in biomedicine, as well as European co-financing of same. They also attempt to connect these issues with the degree of interest in science and technology in general, and in biomedicine in particular (European Commission 2002, 2005a, 2010a). The objective of the *Fifth National Survey on the Social Perception of Science 2010* (*Quinta encuesta nacional sobre percepción social de la ciencia 2010*) carried out by FECYT, was the same as in previous editions in 2002, 2004, 2006 and 2008, that is, to determine how Spanish citizens perceive science and technology. Personal interviews were conducted with 7,744 people of both sexes who had resided in Spain for five or more years and were 15 years of age or more. The field work was done from May 17 to July 9 of 2010, throughout the national territory.[1]

In the case of the Eurobarometer, people were asked, on a Likert-type scale where they had to select two responses, their degree of interest regarding certain topics. Subjects of interest were: nature and environment (84%), medical and health investigation (71%), European and international news (70%), economic and social subjects (68%), sports and outdoor activities (66%), science and technology (60%), art and literature (52%), celebrities and entertainment (42%). The analysis of these results by country shows that 62% of the Spanish population is interested in medical and health investigation, and 50% in science and technology. In spite of dropping approximately 10 points in respect to the European average, the numbers are much higher than those shown in the third national survey done by FECYT on the perception of science and technology, where scientific and technological subjects occupy a modest position on the scale of informational interest in the Spanish population (see Graph 1). Ten percent of those surveyed cite these subjects as being of informative interest to them. It is a similar percentage to that produced by disparate subjects such as terrorism or travel, but remains far below the subjects that lead the chart, such as sports (30%), medicine and health (26%), and cinema and live shows (20%).

The imbalance of data between the European survey and the Spanish one can partly be explained by comparing the response options in both. In the Eurobarometer, the survey-

[1] The sampling error is from +/- 1.14% for a reliability level of 95.5%.

taker is obliged to take a position in each item; however in the Spanish survey they must choose three topics from a much wider variety than in the European survey.

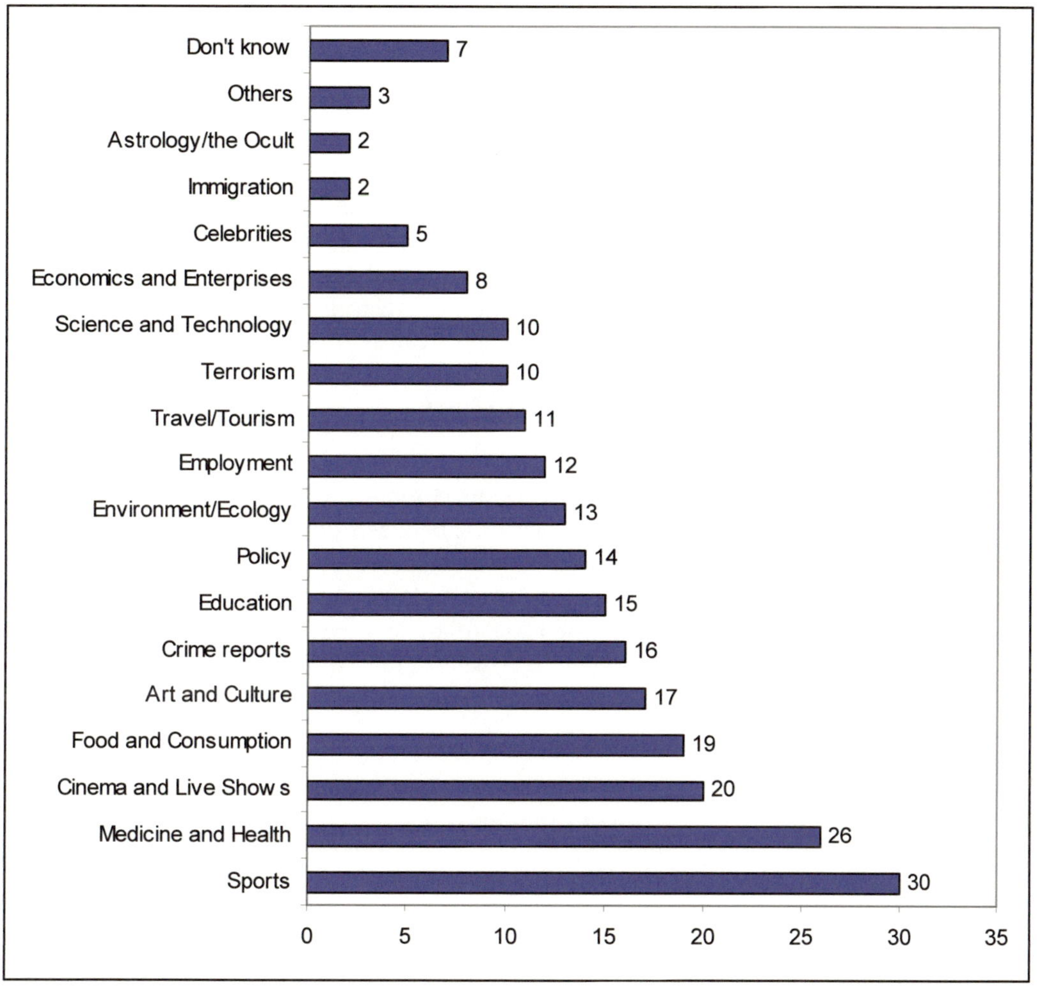

Graph 1. Informational subjects that interest you (máximum 3 responses) (FECYT , 2006)

Among citizens, interest in Medicine and Health (26%) is constant, placing third as a subject of interest in FECYT 2010, behind Work and Employment (32%) and Sports (30%). Regarding the last two, interest may have been influenced by the economic situation, in the first case, and the playing of the World Cup right when the field work was being carried out, in the second. Fifty three point six (53.6) percent of the people are fairly or very interested in medicine and health, although only 29.6% consider themselves fairly or well-informed on these subjects. By subjects of interest, citizens place medicine and health as the subject of most interest, with 3.58 on a scale of 1 to 5. When asked on the survey about which two areas of investigation should take preference in the future, 78.3% of citizens believe that future investigative efforts should center mainly on medicine and health.

To a large degree, all the segments analyzed cite medicine and health as the priority sector in which to concentrate investigative efforts, although it is cited more by women (82.5%). When dealing with questions related to science and technology, the institutions that inspire the most trust (on a scale of 1 "very little" to 5 "much") are Hospitals (4.16) and Universities (4.07). Next come Public Organizations for Investigation (3.79) and Professional Colleges (3.75).

3. Hopes and fears surrounding biotechnology

It's interesting to confirm the importance of dividing the temporal period of perception into before and after 1999. With the exception of Holland and Germany, the majority of European countries tend to maintain low levels of optimism regarding biotechnology, or show a drop in it. Nevertheless, with the exception of Austria, all European countries tend to maintain optimism about biotechnology in the second period, especially in Spain at 75 points, with Malta at the maximum with 81 and Greece at the minimum with 19. As such, the given explanations must offer an account of both movements, without having to be exclusively at the national level (European Commission, 2005a).

This inflection also marks a change in stages of development: from Science Literacy during the period of 1960 - 1980 and Public Understanding between 1985 and the 1990s, to Science-in-Society from the 1990s to the present. The changes in these stages have come with controversies that demand a revision of agendas and discourses. Specifically, the Science-in-Society period emerges from the lack of trust following crises such as the controversy around genetically-modified food. As such, activists begin to proliferate whose analysis and investigations are their way of intervening in technological processes. The idea is to change scientific policy.

Period	Attribution diagnosis	Strategy research
Science literacy 1960s-1980s	Public deficit knowledge	Literacy education
Public understanding 1985s-1990s	Public deficit attitudes	Know x attitude Attitude change Education Public relations
Science-in-Society 1990s-present	Trust deficit Expert deficit Notions of public Crisis of confidence	'Angels' mediators Impact evaluation

Table 1. Paradigms, Problems and Solutions (Bauer, 2009)

One of the ways to explain this change is the concern about genetically-modified food in the first period, and the explosion in biomedical investigation, with its promise to cure illnesses, like the case of investigation into stem cells, in the second period.

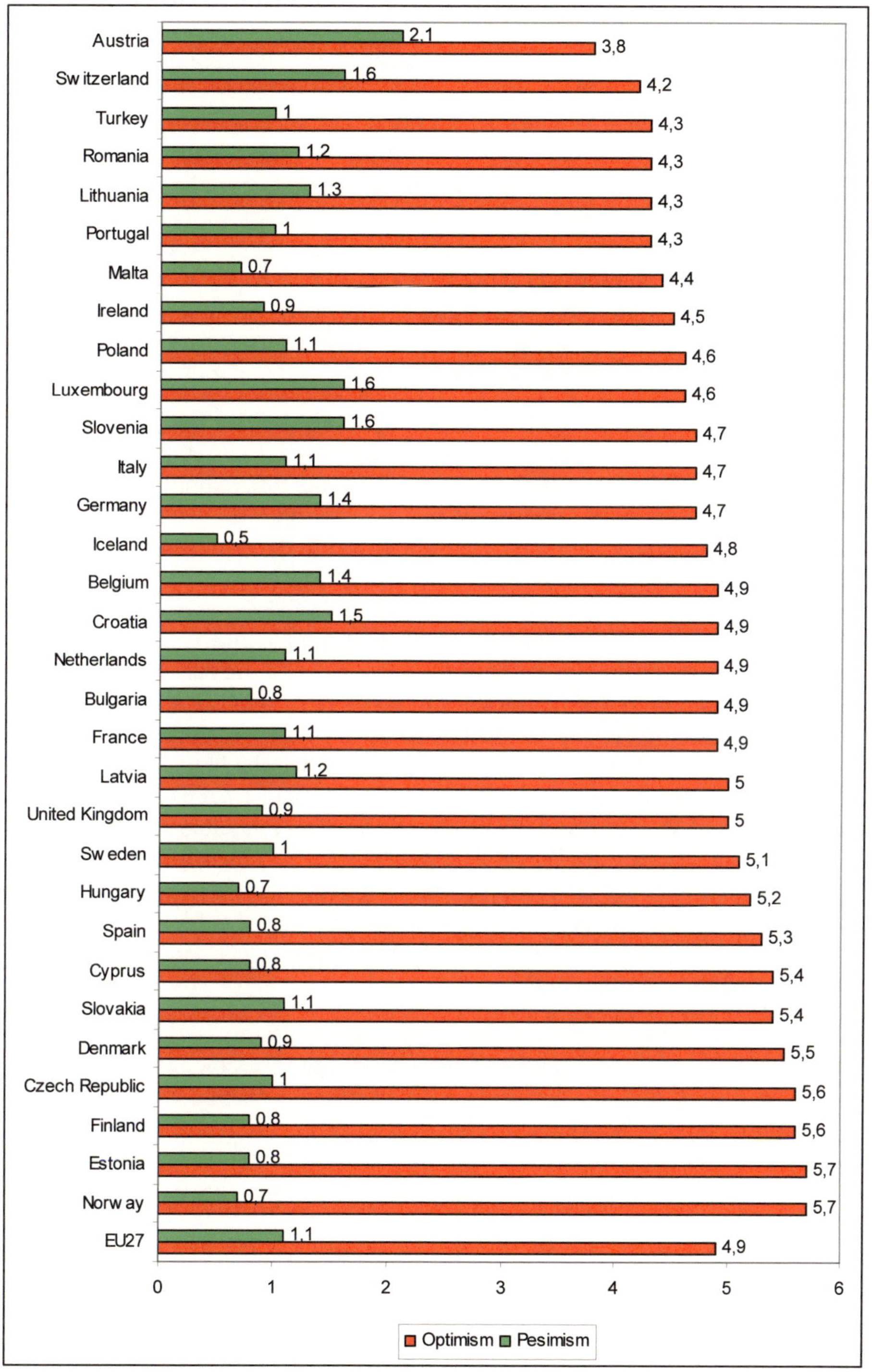

Graph 2. Generalised technological optimism and pessimism. Maximum: 8 (European Commission, 2010a)

3.1 The genetically-modified food controversy

During the decade of the 90s, the image of biotechnology was changing. On one side, the controversy on the distribution of genetically-modified food began. One of the milestones in the debate was importations of genetically modified soy, such as Roundup Ready soy from Monsanto in 1996. The analysis and contrast of different studies made at the European and national level on public perception of biotechnology has been an object of study of previous researches. The ambivalence reflected in the work of Luján & Todt (2000) lies in the difference in the response to moral evaluations of a general type and the attitudes maintained towards specific products. This responds in part to the fact that at the center of the controversy on the regulation of genetic engineering are questions such as if the processes or products are the same or not, if they are substantially or functionally equivalents, and if the techniques are old or new.

The first results of the Eurobarometers in the 1990s celebrated the idea that the public perceived distinctions between diverse biotechnologies. At the same time, regulations throughout Europe were distinguishing between agri-foodstuff technologies and biomedical applications. The regulation of foodstuffs and pharmaceutical products has led to the establishment of procedures which are institutionalized in diverse organizations, unlike the process in the United States with its Food and Drug Administration. Thus, the European Food Safety Authority (EFSA) was set up in January 2002, following a series of food crises in the late 1990s, as an independent source of scientific advice and communication on risks associated with the food chain. For its part, the European Medicines Agency is responsible for the scientific evaluation of medicines developed by pharmaceutical companies for use in the European Union. In Spain, the Spanish Food Safety and Nutrition Agency (AESAN in Spanish) was created in 2001 with the mission to "guarantee the highest level of food safety as a basic aspect of public health and to promote the health of citizens so that they have full confidence in the food they eat and adequate information available to be able to choose[2]. Whereas the Spanish Agency for Medicines and Health Products (AEMPS in Spanish) is "responsible for guaranteeing for society, from a public service perspective, the quality, safety, efficiency and correct information of medicines and health products, from their investigation to their utilization, in the interest of protecting and promoting the health of people, animals and the environment.[3]"

On the other hand, the allocation of funds for investigation is delegated to different organisms that project diverse images of biotechnology. This recognition of the distinct attitudes of the public and of the diverse institutions involved in these technologies contrasts with any type of unifying discourse that can be found in other fields. Or, for example, the declared support of the majority of governments for biotechnology as a "strategic technology for the 21st century." Or when they speak from a sociological standpoint of a "genetic revolution"or of "bioeconomics." That is, phases where it is biotechnologies that would define the future of our societies. One of the umbrella concepts under which diverse interests await is that of "life sciences." It's a concept promoted to cover the university scientific community as well as small companies and

[2] http://www.aesan.msc.es/AESAN/web/sobre_aesan/sobre_aesan.shtml. Last consulted in September 2011.

[3] http://www.aemps.gob.es. Last consulted in September 2011.

pharmaceutical corporations, feigning a certain control over all of them at once. Or even Monsanto, which appears to want to encapsulate an integrated vision of chemical activities, seeds and pharmaceutical activities united with the farmer. Something which favors investors, that is, having a unified image of the technological process and its products (Bauer, 2002).

Quantitative studies showed the ambivalence of the perception in Spain. In them it can be confirmed how Spaniards maintain a positive image of science and technology and the professionals involved in these areas (European Commission, 1993; Moreno, 1996). In the Eurobarometers, these types of perceptions are also maintained (Marlier, 1991). It is the same in comparative analyses, where scientific work is considered fundamental to improve the conditions of life (Bauer & Schoon, 1995).

But one of the principal points of ambivalence detected in this first period refers to the consumption of genetically-modified food. Despite the optimism shown toward science and technology, more than 76% of Spaniards reject genetic engineering, and 72.1% are against its application for human consumption. One of the reasons can be found in the gastronomic culture, which doesn't much accept industrially treated foodstuffs. Also influential were the controversies over poisoning from the ingestion of colza oil not suitable for human consumption. On the other hand, in this period of analysis, the Eurobarometer never asked about the evaluation of the application on different organisms, nor about obtaining different products. Rather, it described a generic type of investigation by way of some examples and questions about the level of agreement or disagreement with affirmations like "investigation (on plants, microorganisms and farm animals) should be supported." That is, they asked about the processes and not about the products of investigation. Moreover, when the question was about support for applications of biotechnology and genetic engineering on different types of organisms, a moral evaluation was made; and when the question was about a specific application, its utility was evaluated. Hence, the ambivalence is the result of the conflict between evaluation of the process and evaluation of the product. So, while the application of genetic engineering on plants and foods received a positive moral evaluation, the usefulness of the products obtained from these processes lacked support.

"Question (Q): Do you agree that genetic manipulation should be used for. . . ?"

	"Yes"	"No"
Diagnoses for hereditary diseases	96.2	3.8
Herbicide resistant plants	69.5	30.5
New gene therapies	66.6	33.4
Bigger fish for consumption	27.8	72.2
Faster growing livestock	23.8	76.2

Unit: percentages.

Table 2. Attitudes towards applications of genetic engineering (IESA, 1990 in Luján and Todt, 2000)

"Q: Nowadays it is possible, for example, to introduce genes of corn in potatoes to increase their nutritional value. Would you be willing to consume this type of potatoes?"

Yes	39.6
No	48.3
DK/NA	12.1

Unit: percentages.

Table 3. Attitude towards the consumption of transgenic products (Atienza & Luján, 1997)

"Q: Please tell me whether you tend to agree or tend to disagree with the following statement: I would buy genetically modified fruit if it tasted better."

Tend to agree	28
Tend to disagree	50
DK/NA	22

Unit: percentages.

Table 4. Attitude towards the consumption of transgenic products II (European Commission, 1996)

3.2 The promise of stem cell research

Biomedical investigation has had more support than agricultural applications of biotechnology, and has a huge social and economic impact. Its hopeful aspect as a cure for illnesses and improving people's quality of life has been influential in this. However, it appears that regenerative medicine, specifically, also arouses controversies and social, political and religious tensions.

As we see in Graph 3, levels of acceptation are high. Around 68% are favorable towards investigation with non-embryonic stem cells and 63% for the use of embryonic stem cells. However, it's not unconditional support. It varies by country and depends on the legislation and control that governments exercise over the investigations. It is limited by the subordination of research to its respective legislation and controls. In fact, approval percentages under strict laws reach 44% for human enhancement and 54% for investigations with non-embryonic stem cells. Thus, governance is essential in these investigations. And as such, regulation of stem cell investigation varies in European countries. A striking aspect of these investigations on an international level is the diversity they have achieved in a short period of time. There are many countries, like Israel, Sweden and Singapore, that concentrate their economic efforts in specific niches of these technologies. Others, like China and Korea, lacking experience in biomedical investigation and development, have constructed investigative installations that are competitive. Nevertheless, other countries like the United States, Germany and Australia, among others, have seen their investigations hindered due to religious and political debates on the use of human embryos for investigation, therapeutic cloning and the generation of hybridoma cells.

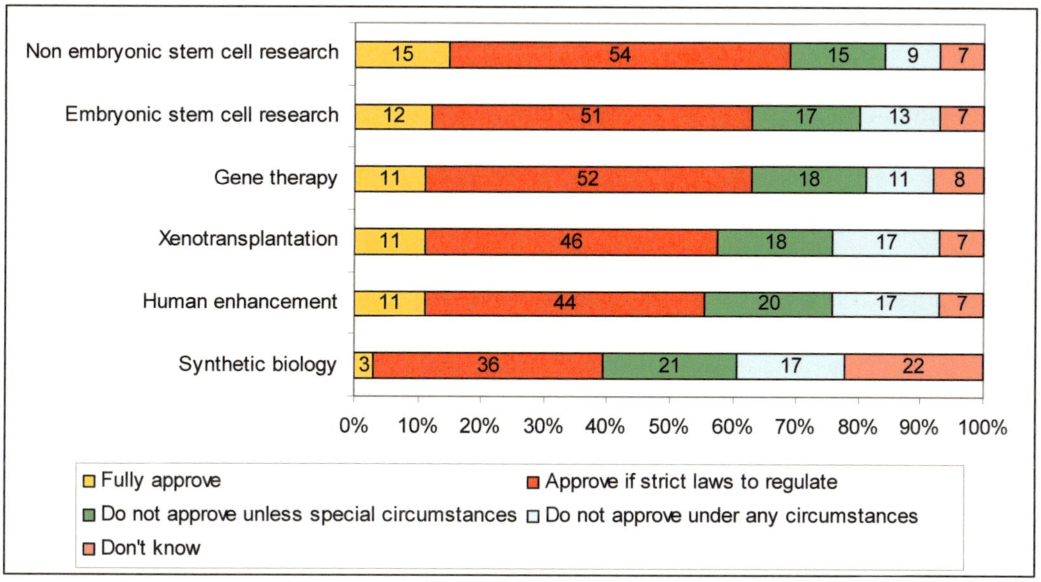

Graph 3. Levels of approval of biomedical research and synthetic biology, EU 27 (European Commission, 2010b)

In fact, although research with stem cells (embryonic or adult) appears to be an important source of knowledge to improve therapies and fight against illnesses, moral aspects exist that determine the support of some citizens for embryonic stem cell research. The policies of scientific popularization assume that a higher level of scientific knowledge among citizens can help in having greater support for investigation. In fact, some studies have shown that the opinion that citizens have about certain controversial investigative subjects, as is the case with stem cells (SC), isn't influenced by the information they receive but rather by moral or religious aspects (Nisbet, 2005). Nor does it seem correct to assume that a higher level of scientific knowledge is associated with more support for SC research. It's possible that individuals who are more informed have a firmer position, whether in favor or against SCs (Bauer, Allum y Miller, 2007).

In Spain, there is an average understanding of the properties of embryonic stem cells. According to the BBVA study, 2009, 41.9% of those surveyed knew that stem cells can be transformed into various types of different cells and change into specialized tissue like muscles or nerves, but didn't know that the extraction implies the destruction of embryos. Spain (along with the Czech Republic) is the country that most supports research with embryonic stem cells that are only a few days old, in hopes of finding effective treatments as soon as possible for illnesses such as Parkinson's, Alzheimer's or diabetes, at 6.8 (on a scale of 0 - 10 where 0 means complete disagreement and 10 means complete agreement). This can be interpreted to mean that acceptance increases when medical benefits are specified in possible treatments for illnesses that citizens consider important. And this is consistent with Spain being one of the countries that least considers research on days-old embryos as being an unacceptable interference in the natural processes of life (4.9; only Denmark is lower, 4.7).

Spain is also one of the countries where research with embryonic stem cells is considered very useful (6.4 on a scale of 1-10, only superseded by Denmark and Sweden, 7.0) and isn't imbued with important risks (5.0), superseded only by Denmark (5.5), Holland (5.2) and the Czech Republic (5.1). The idea of usefulness is quite established, although this doesn't suppose the disappearance of the perception of risk or immorality. Only in certain countries (such as Austria, Germany and Poland) is immorality a more significant feature than usefulness, according to BBVA, 2009.

The financial restrictions on embryonic stem cell research in the United States may have inspired other countries, many of them in Asia, to promote this investigation through specific initiatives for regulations and financing. In spite of the limitations on federal financing, the United States has developed good alternatives such as, for example, through industry or philanthropic investment. In Europe, nevertheless, many countries show strong support for stem cell research. A good example of this is the United Kingdom and their efforts to create transparent policies. Their main installations are located in London, Cambridge and Edinburgh. Sweden has developed dozens of cell lines, carrying out the first studies on the transplant of fetal cells in the treatment of Parkinson's disease, thus stimulating the use of these cells in the treatment of neurodegenerative disorders. Germany, which has legal barriers on research done with embryonic stem cells, has established excellent centers in Berlin and Munich for the study of somatic cells and their potential use in regenerative medicine.

Cooperation and diffusion among countries is one of the characteristics of this research at the European level. The European Science Foundation launched a program, *EuroStells*, for the analysis of comparative research between stem cells obtained from different sources. The *Sixth Framework Programme* backed the development of a database for embryonic stem cells. Currently, scientists and communicators from 90 laboratories are gathered for the *VIIth Framework Programme* to do collaborative work. Nevertheless, European policies haven't managed to smooth over the different policies between countries.

The governments of many countries in Asia and Oceania have demonstrated extraordinary support for the development of stem cells and their application. China, Korea, Singapore, India and Taiwan have invested in this research since 2001. Japan and Australia have built large foundations in basic biology and clinical development in order to create the main stem cell institutes in Kyoto, Kobe and Melbourne. In 2007, a research network was created between Asia and the Pacific (SNAP), launched by scientists in eight countries, but without reaching competitive levels. At the national level, many of the Asian countries have organized strong national societies, which is the case in Singapore, Taiwan and Korea.

Graph 4 shows the level of support that embryonic stem cell research had between 2005 and 2010. Around 55% of people in 19 countries support this research. Support has risen eight points or more in Estonia, Finland, Greece, Ireland, Latvia and Slovenia. On the other side, it has gone down another eight points or more in Hungary, Italy, Poland, Cyprus, the Czech Republic, Germany, Slovakia and Austria. As such, the graph points to future controversies.

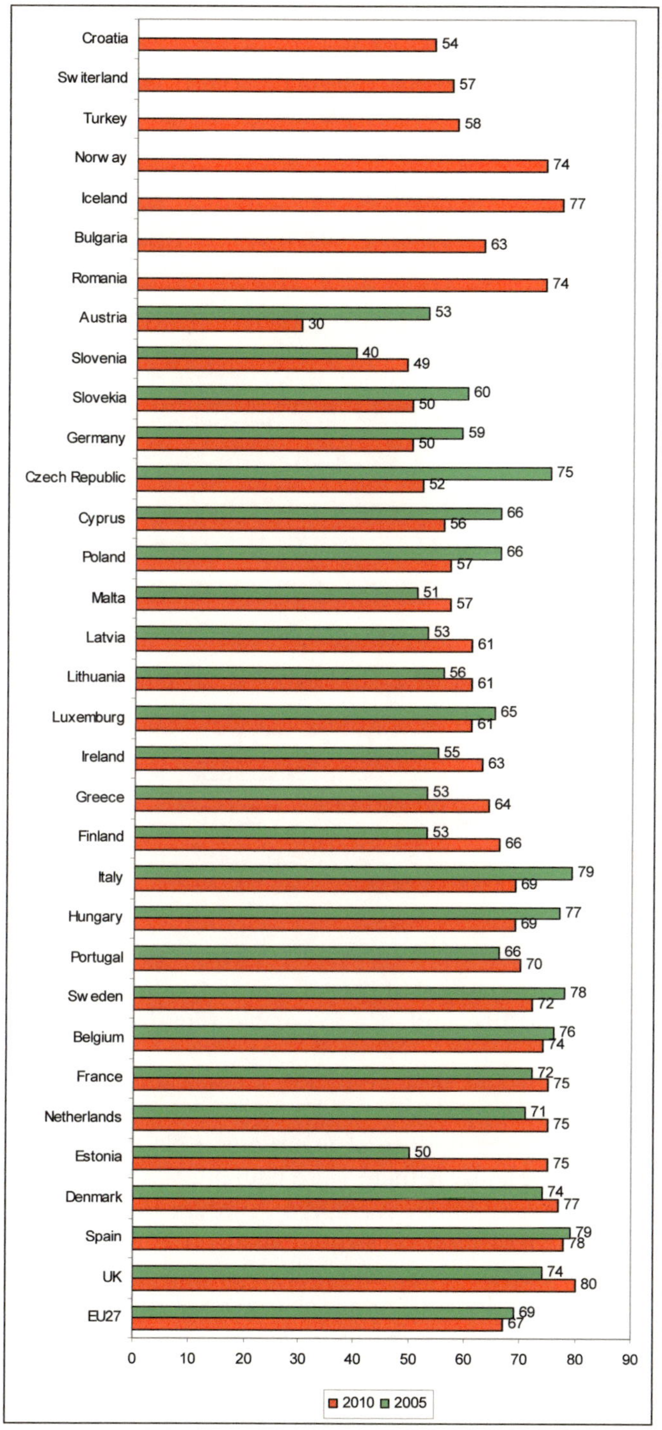

Graph 4. Levels of approval for human embryonic stem cell research, 2005 and 2010. % of respondents (European Commission, 2010b)

One of the factors not included in the Eurobarometer, but adds clues to the perception of this type of research, is the evaluation of the use of induced pluripotent stem cells (iPS), which would avoid the use of embryonic cells and, as such, close many debates. According to the study by the Fundación BBVA, acceptation increases in the case of techniques that don't harm or destroy the embryo.

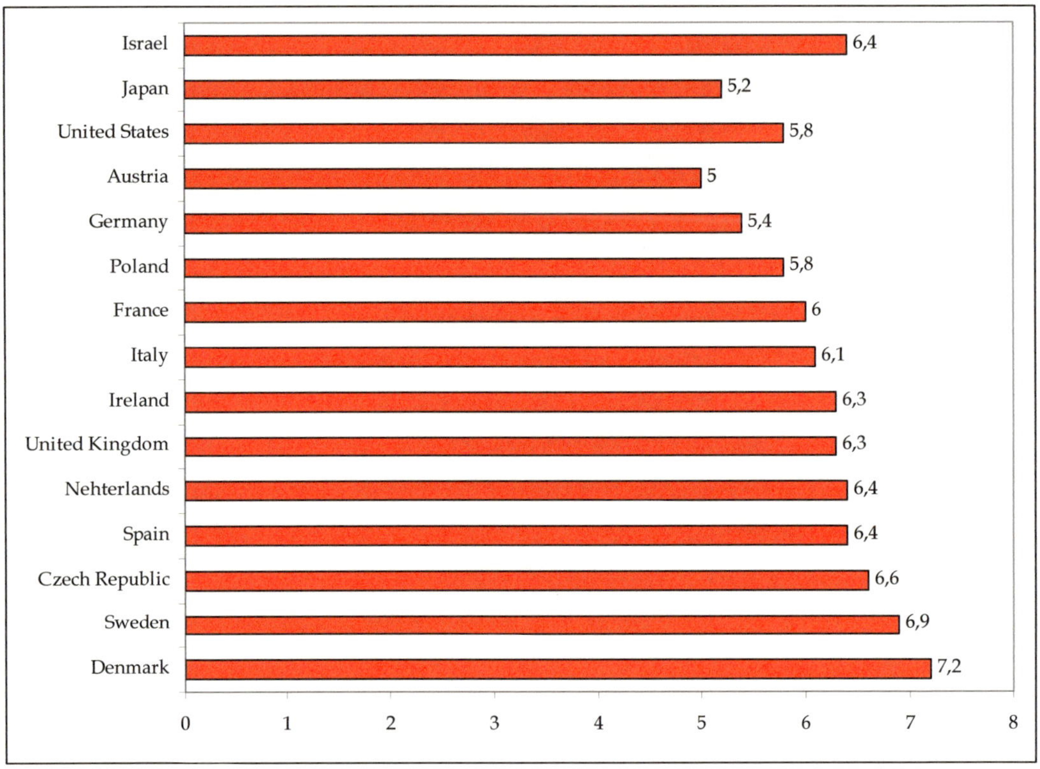

Graph 5. Situation without destruction of the embryo. 0 signifies completely unacceptable from a moral perspective and 10 completely acceptable. (Fundación BBVA, 2008)

The polarization of political and public opinion around the ethical questions related to research with embryonic stem cells has contributed to the growth in interest in obtaining adult stem cells. The defense of the use of embryonic cells tends to ride on its potential future use for cellular transplants. If there were equivalent alternatives, it would be more complicated to defend the use of human embryos in order to get stem cells from them. In this sense, induced pluripotent stem cells could change the terrain of the ethical debates. Those opposed to the use of embryonic stem cells might interpret the results of the research done with stem cells (adult or derived from umbilical cord blood) in a more optimistic way than the experiments show. In the same way, defenders of human embryonic stem cells might be less open if the potential risks of these cells in clinical use were considered. Private clinics operate precisely in this confused space, confronting a medical necessity that millions of patients with incurable diseases demand. Commercials for stem cell clinics often represent the patients as being responsible for their own destiny, while portraying standard clinical medicine as reactionary. They appeal strongly to the needs of patients, some with chronic illnesses or disabilities, whose current medical treatment doesn't offer a solution. As a result, patients and their families spend many hours surfing the internet looking for stem cell treatments for their particular illness. They're attracted by advertisements on the internet, blogs, articles in national and local magazines and other press, while treatments are publicized as having been successful in Mexico, Russia, India, China, Africa and other countries. Often they'll find not one, but dozens of options, generally in countries where there isn't legislation that controls the activity. Clinics also know how to change their name and country quite rapidly if there's an attempt to discredit them. A patient's secondary effects are often difficult to prove, patients are reluctant to show they've gotten worse, and they rarely admit that they have not benefited from the treatment. One of the reasons is the feeling they have of having paid a huge sum of money for something that hasn't worked, or also the placebo effect that for some ailments can lead to significant, if temporary, improvements.

3.3 A future third period?

The report from 2010 seems to open a third period of falling optimism between 2005 and 2010, at least by one point, as can be seen in Graph 6 (European Commission, 2010). We can continue talking about optimism, but the changes that arise in biotechnology and genetic engineering, and the tendencies the surveys are currently showing, is what especially interests us. If we pay attention to Graph 6, the rise in optimism for wind, solar and nuclear energy in this last period contrasts with the fall in optimism for Computers and Information

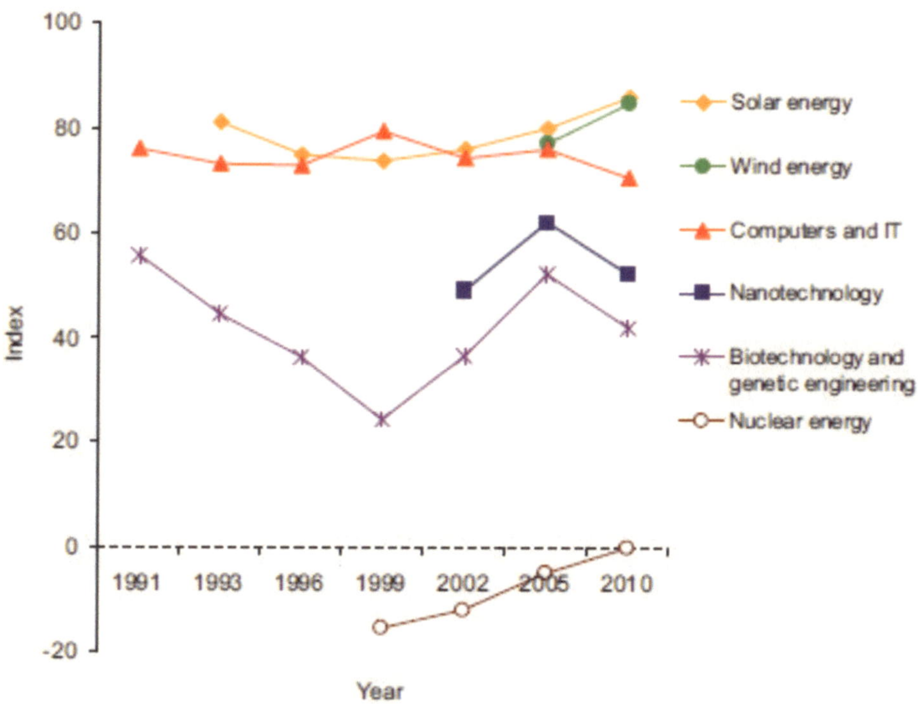

Graph 6. Trends in the optimism index of certain technologies (European Commission, 2010a)

Technologies (IT), nanotechnology, biotechnology and genetic engineering. On one side, alternative energies show a rise because of the "Copenhagen Effect." The controversy surrounding climate change, global warming and carbon emissions has helped feed optimism in renewable energies. Regarding the rise in nuclear energy, it must be taken into account that while those who support solar energy also support wind energy, their opinions are divided regarding nuclear energy, between optimism (46%) and pessimism (42%).

How to explain, then, the drop for computers and IT, nanotechnology, biotechnology and genetic engineering? The 2010 report considers that nanotechnology as well as biotechnology show this drop because of the rise (from 12% to 20% in biotechnology) in the response "make things worse." The disenchantment society feels seeing highly promising announcements that later don't materialize in products, therapies, etc., has a lot to do with it. The possibility for future conflicts, especially between the Spanish food industry and consumers, derives from the gap between the perception of the industry toward consumers, mainly as being passive, and the attitudes of the consumer (Todt et. al., 2009).

4. Conclusions

Biotechnology constitutes a good indicator of the governance of science and technology. For the plurality of its practices, periods can be analyzed to see which show distrust, optimism, fear, and in function of which technologies. The main ambivalences found in this study are:

- European and Spanish indicators show certain divergences in respect to certain processes, as in the case of the perception of science and technology. The lack of homogenization in the surveys, their units of measure and objectives make a comparative analysis of same difficult. The divergence in the results makes it necessary to pay attention not only to the analytical differences they present, but also the qualitative aspects of the processes analyzed.
- In spite of the political, economic and media dominance of unifying discourses in biomedicine (i.e.: life sciences, genetic revolution, bioeconomics), the public shows a complex and plural perception regarding these processes. Positioning and evaluating the different biotechnologies analyzed such as genetic engineering, regenerative medicine, cellular therapy, etc., and modifying their perceptions, attitudes and actions in function of the course taken by them. It is precisely this flexible aspect that obliges institutions to maintain transparent policies in which citizens keep taking an ever more active part in the decision-making.
- A moral evaluation on biomedical processes and the utility of their products continues to be posed. The case of the perception of stem cell research is a clear example of this. The positive evaluation of research with induced pluripotent stem cells, or even with adult cells, in contrast to research with embryonic cells shows a moral reflection about these processes. That is, that the use of the products from these technologies, as well as the hope placed in them to cure illnesses, is accepted independent of the moral evaluation of said technologies.

To sum up, analysis of the processes of the perception of biotechnology put into question the homogenization of national and European tools when approaching the same questions. At the same time, diversity, the capacity for change and the complexity of the perception of citizens about these biotechnologies requires giving incentives to European and national mechanisms to favor greater public participation in the regulation and risk prevention of them.

5. Acknowledgment

This chapter was carried out as part of the Cartographies of Science and Technology: Ethnographies, Images and Epistemologies (FFI 2009-07138-FISO) project.

6. References

Atienza, J. Luján, J.L. (1997), *La Imagen Social de las Nuevas Tecnologías Biológicas en España*, CIS: Madrid (*The Social Image of New Biological Technologies in Spain*)

Bauer, M. W., (2009), "The evolution of public understanding of science-Discourse and comparative evidence", *Science, Technology & Society*, 14: 221 - 242

Bauer, M. W., (2003), "Controversial medical and agri-food biotechnology: a cultivation analysis", *Public Understanding of Science*, 11: 93 - 113

Bauer, M., y Schoon, I., (1993), "Mapping Variety in Public Understanding of Science" *Public Understanding of Science*, 2 (1993): 141-155

European Commission, (1991), *Eurobarometer 35.1.* Commission of the European Communities: Brussels

European Commission, (1993), *Eurobarometer 38.1. Europeans, Science and Technology: Public Understanding and/ Attitudes*, Commission of the European Communities: Brussels

European Commission, (1996), *Eurobarometer 46.1.* Commision of the European Communities: Brussels

European Commission, (2002), *Eurobarometer 58.0. Europeans and Biotechnology in 2002*, European Commission.

European Commision, (2005a), *Eurobarometer 64.3. Europeans and Biotechnology in 2005: Patterns and Trends*, European Commission: Luxembourg

European Commission, (2005b), *Europeans, Science and Technology*, European Commission: Luxembourg

European Commission, (2010a), *Europeans and Biotechnology in 2010. Winds of change?*, European Commission: Luxembourg

European Commission, (2010b), *Science and Technology*, European Commission: Luxembourg

FECYT (2006): *Tercera encuesta nacional sobre percepción social de la ciencia*, Madrid.

FECYT (2010): *Quinta encuesta nacional sobre percepción social de la ciencia*, Madrid.

Fundación BBVA, (2008), "Actitudes hacia la investigación con células madre", *II Estudio de Biotecnología de la Fundación BBVA* ("Attitudes Toward Stem Cell Research", *II Study of Biotechnology from the BBVA Foundation*)

IESA, (1990) *Biotecnología y Opinión Pública en España*, Instituto de Estudios Sociales Avanzados: Madrid (*Biotechnology and Public Opinion in Spain*, Institute of Advanced Social Studies: Madrid)

Luján, J. L. y Todt, O., (2000), "Perceptions, attitudes and ethical valuations: the ambivalence of the public image of biotechnology in Spain", *Public Understanding of Science*, 9: 383-392

Marlier, E., (1991), "Eurobarometer 35.1: Opinions of Europeans on Biotechnology"in *Biotechnology in Public*

Moreno, L., (1996), "La Opinión Pública y los Avances en Genética," in ed. Borrillo, D., *Genes en el Estrado,* CSIC: Madrid ("Public Opinion and Genetic Advances")
Pérez Sedeño, E. And Miranda Suárez, M.J. (2008): "Percepción pública de la biomedicina en España", *Medicina Clínica,* : 131 (Sup. 3) : 6-12 ("Public Perception of Biomedicine in Spain," *Clinical Medicine,* : 131 Sup.3)
Todt, O., Muñoz, E., González, M., Ponce, G., Estévez, B., (2009), "Consumer attitudes and the governance of food safety", *Public Understanding of Science,* 18: 103-116